W9-DAN-897

The Energy Balance Diet

The Energy Balance Diet

Joshua Rosenthal and Tom Monte

This book is dedicated to improving the world's health and happiness through better nourishment of all people.

International Standard Book Number: 0-02-864358-5
Library of Congress Catalog Card Number: 2002110193

04 03 02 8 7 6 5 4 3 2 1

Interpretation of the printing code: The rightmost number of the first series of numbers is the year of the book's printing; the rightmost number of the second series of numbers is the number of the book's printing. For example, a printing code of 02-1 shows that the first printing occurred in 2002.

Printed in the United States of America

For marketing and publicity, please call: 317-581-3722

The publisher offers discounts on this book when ordered in quantity for bulk purchases and special sales.

For sales within the United States, please contact: Corporate and Government Sales, 1-800-382-3419 or corpsales@pearsontechgroup.com

Outside the United States, please contact: International Sales, 317-581-3793 or international@pearsontechgroup.com

Contents

Appendixes

Introduction

The Only Diet That Can Work for You Is the One You Create

Animals in their natural habitat instinctively know exactly what to eat. They are aware of which foods are appropriate for them and which foods are not. Nature has given them an innate wisdom. They do not need courses on nutrition, or diet books, or scientists to tell them what to eat or how to eat. They just know. And they are remarkably healthy: no obesity, depression, hay fever, or cancer. They don't need doctors, hospitals, HMOs, or insurance. And if they get out of balance, they instinctively know what to do to make themselves well again.

It's highly unlikely that nature gave this wisdom to wild animals and not to humans. Deep inside of us, we know what's good for us and what's not. Somewhere along the line, we got separated from the wisdom that's in every one of us. Or, rather, our innate knowing got buried under so much "education" from teachers and experts.

In order to appeal to the widest possible audience, diet proponents have to convince people that they have the diet that's right for everyone. If you read a book by someone selling high-protein diets, guess what that book is going to urge you—and everybody else—to eat? Lots of protein! If you have type O blood and read a book by an advocate of the blood type diet, that book is going to urge you to adopt the diet for type O people—or for whatever blood type you are. If you read a book by an expert on vegetarianism, the author will try to convince you to eat only vegetables. But there are so many variables that determine what you should eat that it is impossible for someone to create a diet that's right for you under all circumstances, never mind that it would not be right for everyone else as well.

Think about it: Should a construction worker and a secretary eat the same diet? What about a young woman in her 20s and a man in his 60s? Should they both eat the same high-protein diet, or the same blood type diet, even if they both have "O" blood? Of course not. That would not be practical.

What about the place where a person lives? Should a person who lives in India eat the same diet as someone who lives, say, in Sweden? Can a person who lives in Italy be happy on a Russian diet—or a Russian on an Italian diet? I don't think so.

What about the season? Does your body want the same foods in the winter as it does during the summer? No. How about the day of the week? Don't you eat differently on Saturday night than you do on Tuesday night? Most people do.

If all of these variables matter, then how can we create one diet that will address everyone's dietary needs? We can't. No one can create a diet to fit all environments, all lifestyles, all sizes, shapes, and sexes, all ethnic and racial backgrounds, all ages and circumstances. It isn't possible. You can adopt the latest fad diet with all the discipline and good intentions you can muster, but sooner or later you won't be able to follow it, because your body is going to rebel.

That's the primary reason why diets fail. Something inside you rises up and insists that your current way of eating isn't supporting your biological and psychological needs. The longer you stay on it, the more tense your body becomes. Pretty soon, you start craving foods that you feel will relieve you of your tension and restore balance to your condition. Maybe you fight those cravings for a while, but eventually you give in. And then you find that you're no longer practicing your diet. Like the squirrels, deer, coyotes, and bears, an innate wisdom lives in your cells. You can fight that wisdom, but sooner or later it emerges to tell you that what you'd really like to eat is a big salad or a plate of spaghetti, or, if you repress your natural desires long enough, a large brownie.

The Body Is Rarely—If Ever—Wrong

We have been trained to ignore or deny the body's innate wisdom. We have been taught to believe that the body is wrong. Cravings are a sign of the body misbehaving. We need to discipline ourselves, we are told. But is it really possible that the body is so poorly designed? Is the body really the problem?

The human body is the perfect biocomputer. It is constantly adjusting its biochemistry to adapt to its internal and external circumstances in order to maintain health and survival. Let me give you an example.

During World War II, pregnant Dutch women were forced to live for six months on 450 calories a day, about one sixth of what experts say is adequate for a pregnant woman. The women were starving, and so were their unborn babies. Under those conditions, doctors expected a tremendous increase in stillbirths, and premature, underweight, and even handicapped babies to be born. None of that happened. Researchers, whose findings were later reported in the medical journal *The Lancet* (338:415;1991), discovered that the weight of the average baby born during that period was only eight ounces less than those children born during times when there was plenty of food. There was no rise in infant death rates, nor in physical or mental disabilities. The children were as physically healthy and as smart as babies born at any other time. What happened? The women's bodies adjusted to the starvation diets. They were able to sustain not only their own health, but the health of their babies, too.

The body is always adjusting to its circumstances. The body knows when it's tired, when to fall asleep, when to wake up, when it's hungry, and when it's had enough to eat. It knows when to eliminate waste and when to store nutrition. If you put something poisonous in your mouth—and we all do several times a day—the liver recognizes the toxin and immediately goes into action. It creates just the right antidote to neutralize the substance on the spot. Despite all its heat-producing functions, the body constantly maintains its ideal temperature of 98.6 or thereabouts. There is no greater miracle than the human heart. For the average person, the heart beats 100,000 times a day, and about 2.5 billion times in an average lifetime, and not one of those beats happen because you thought to tell your heart to beat. All the major events inside the body occur without your conscious thought, will, or discipline. You can be working or playing hard and forget all about food. But at some point during the day, your body will intrude upon your fun and you'll find yourself saying "Hey, I'm hungry."

Countless biological functions arise from an intelligence that exists in your cells, tissues, bones, and organs. That intelligence is constantly monitoring the overall workings and condition of the body, making minor and major adjustments, in order to maintain health and life. From that inner knowing, we become aware that we need more rest, that we've got a cold coming on, that our feet hurt, that we're happy, that we're sad, that we're up for the challenge, that we would like to eat Italian food tonight.

Cravings are not random. They arise out of an intelligence that exists inside us, an intelligence that is constantly monitoring our inner condition and attempting to use food and behavior to create balance and health.

The trouble is, we don't understand cravings or their source. Without such an understanding, we can't restore balance to our eating habits. One of the things that makes it impossible to understand cravings, and the tension that accompanies them, is that we have been taught to be afraid of food. Fear of food is really fear of the body. We're afraid that if we eat this or that food—even when we truly desire it—we'll get sick and suffer. Therefore, we must listen to the experts, because they can tell us what to eat to avoid suffering.

But when you listen to the experts, you get a cacophony of conflicting theories. High-protein, low-carb. High-carb, low-protein. Vegetarianism. Macrobiotics. Raw foods. Juicing! Advocates of every one of these diets can cite scientific studies to prove that theirs is the one true way of eating. Diet is the one area of human endeavor where people can scientifically prove that their theory is right and everybody else is wrong.

Most Diets Work—but Not for Long

Here's a fundamental principle that I hold throughout this book: Most diets have truth in them, most diets provide certain benefits, and most proponents of different diets can prove their approach scientifically and anecdotally. In other words, there are always going to be people who experience some benefit from every popular diet that's being marketed today, even if it's something as crazy-sounding as the grapefruit diet. High-protein diets do cause weight loss, and sometimes significant weight loss. There is some scientific support, as well as sound reasoning, behind the blood type diets. People with O blood often do need more animal foods in their diets, for example, than people with B blood. High-carbohydrate, low-fat diets can prevent and reverse most serious disorders, including heart disease, adult-onset diabetes, high blood pressure, overweight, and, in some cases, even cancer.

Diet may be the most effective self-help healing tool available to humans. In fact, there is nothing more powerful for changing your physical and mental health than appropriate diet. The right diet, at the right time, will have a profound—even life-changing—impact on your health.

But here's the problem that proponents of specific diets have: Every one of them overpositions and oversells his program. Each insists that his diet alone is the right way to eat, and the only right way to eat. And that's where everyone runs into trouble. There are times in your life when you need more protein, just as there are times when you should abstain from animal foods entirely. There are times when your body can tolerate more sugar. At those times, a small amount of high-quality chocolate actually may be good for you! There are other times, however, when you should absolutely avoid such foods. In general, most people benefit by avoiding dairy products entirely, but some may benefit from occasionally eating small amounts of high-quality yogurt. But what are small amounts, exactly, and what does "occasionally" mean? Only you can decide if your body benefits from that food, and how much of it you need.

It is extremely difficult, if not impossible, to prescribe the absolutely right quantity of a specific food in a mass-market book. But diet proponents do it all the time. They tell you to avoid carbohydrates, or to stay away from animal foods. You may adopt one or another of these diets with all the sincerity and discipline of a monk. But it doesn't take long before you start craving foods that are considered taboo. The longer you remain on the diet, the more tension and frustration you experience. And when you finally decide you can't resist any longer and give up on the whole thing, you feel like a failure.

Where does that leave you? It leaves you alone with yourself. And that can be a very good place to be.

The Right Program for You

The wisdom to restore your health, lose weight, and feel great is already inside you. In this book, I am going to help you reunite with that wisdom and cultivate it so that you can create a way of living and eating that's right for your body and your overall life. That way of eating can improve your health and give you abundant energy, optimal weight, a more youthful appearance, and protection from major disease. But it can give you many things that you don't typically think of as related to food.

If you eat a diet that someone else describes as ideal, you are a follower of that person. He or she is essentially telling you what to eat. But if you eat a diet that arises from the wisdom that is in you, then you are becoming more yourself with every meal. You are learning to be in contact with the most precious thing you have been given: your inner nature, your life.

In this way, food can be an instrument that brings you closer to yourself and to who you really are. It can give you the freedom to be you.

I am the founder and director of the Institute for Integrative Nutrition in New York City. Each year, hundreds of people from all walks of life attend our professional training program and learn how to create a way of eating and living that's right for them. They also learn how to share that knowledge with others in a professional way. People discover how to create a diet that's in harmony with their constitutions, ages, genders, heritages, body type, lifestyles, and ambitions. They discover that their own unique ways of eating can help them achieve their health goals. They also learn that only this way of eating can be maintained for the rest of their lives.

I also teach people the deeper importance of food and eating, which I will also present in this book. I want to reveal not only how you can find your way to health, but what health can mean for you.

I came upon my approach through many years of diligent study and by working with the dietary patterns of thousands of people. I also adopted many programs being offered today. In the end, I found that most current nutritional approaches provide some benefit. Every diet works for someone, at least to some extent, and at least for a short period of time. But like so many people today, I also found that I could not follow any particular program for very long, because none of these diets could adapt to my ever-changing dietary and nutritional needs.

When I say that to people, they immediately think that I am advocating self-indulgence and overeating. But the intelligence that is woven into your cells does not want you to overeat. You suffer when there is too much food in your stomach. Your inner wisdom doesn't want you to eat excessive amounts of fat, because your heart suffers and your body has to labor in order to remain alive. Your inner wisdom doesn't want you to eat too much sugar or processed foods, because they rob your body of nutrition and cause you to gain excess weight. The intelligence in your cells does not want you to be overweight or excessively thin, because both conditions make it more likely that you will become seriously ill and die prematurely. Your inner wisdom wants you to be healthy, vital, energetic, free of excess weight, clear of mind, and joyful of heart. What we have to do is to start listening to the body in order to know what it needs to be happy.

Here's a simple truth about the human body: Given half a chance, it will heal itself of almost anything. If you reduce meat, milk products, sugar, coffee, alcohol, tobacco, and chemicalized and artificial food, I guarantee that you will feel happier and healthier within a few days. If you add more sleep and clean, pure water to that simple regimen, you'll feel even better. The human body is constantly striving to restore its health and youthful vitality. It does not want to die before its time.

Humans are designed to be healthy. But the key to good health is finding the way of eating that supports you biologically, psychologically, and spiritually. Only a way of eating that comes from your inner knowing can satisfy you in all three of these ways. My intention with this book is to show you how to find that approach and to help you enter that state of wisdom that is already within you. Allow me to start by telling you how I came to this way of eating and living myself.

My Own Path: Learning Where to Find the Answers

Beginning in the 1970s, I began to search for ways to improve my health through dietary change. One of the first programs I adopted was vegetarianism. The people I met who practiced vegetarianism ate no meat, no sugar, and no processed foods. Instead, they ate grains and vegetables and foods with strange-sounding names like tofu, tempeh, hummus, seitan, arame, and nori. They made their own yogurt because they said most yogurts sold in stores were dead, whatever that meant. I didn't want to ask what they meant because I didn't want to look stupid. Later I read that the bacteria and enzymes in most store-bought yogurt were destroyed by pasteurization.

Vegetarianism led to a raw-foods diet. The proponents of raw foods argued that cooking destroyed the nutrition and enzymes in foods. While I was on a raw-foods diet, I encountered macrobiotics, which offered theories that were very different from those of raw foods. Both were plant-based diets, and both agreed that meat was unhealthy, but the raw foods people were adamant that nothing should be cooked. The macrobiotic people were equally dogmatic that everything should be cooked. Hmmm … what should I do? I wondered.

It was wintertime in Canada when this conflict arose. It was cold outside. Snow was falling and the wind was blowing, and my body told me that warm, cooked food was far more appealing than food that was cold and uncooked. I listened to my body and ate the cooked foods.

Macrobiotics, which is based to a great extent on traditional Chinese medicine (TCM), gave me the priceless gift of understanding the body's constant need for balance. As with the principles of TCM, macrobiotics taught that food cravings are usually the body's way of making balance with the effects of previously eaten foods, as well as the person's lifestyle. For example, people who eat a lot of cooling foods, such as papaya and other tropical fruits, oftentimes crave warming foods, such as pizza, spices, and animal foods. People who eat salt, hard cheeses, and lots of red meat, all very contracting and tension-producing, crave foods that make the body feel more relaxed and expanded, such as sugar, processed foods, soft drinks, and alcohol. The more you ate of the contracting foods, the more you craved the expanding foods, and vice versa.

Even lifestyle can create imbalances that result in cravings. For example, if you work hard all day, or all week, or all month, you very likely will find yourself craving foods that create a more relaxed state, such as vegetables and pasta, or—more extreme foods—such as sugar and alcohol. People find themselves craving foods and behaviors that balance their condition. Macrobiotic philosophy teaches that if we eat a more balanced diet, and live in a more balanced lifestyle, cravings for foods that have an extreme effect on the body gradually fade away.

Eventually, I discovered that cravings could arise from even deeper sources than imbalance. My parents and grandparents come from Hungary and, like them, I grew up eating meat and dairy foods. Macrobiotics was a very different way of eating from the diet I grew up on. Maybe it was just a coincidence, but every so often, while I was practicing macrobiotics, I found myself craving meat. It was as if my whole body was salivating whenever I smelled meat being cooked. For a long time, I resisted. Macrobiotics gave me the belief that I had reached a kind of enlightened purity. My desire for meat did not subside, however. It got stronger. I even started dreaming about eating meat. So I did.

Every so often, I would find that my car had pulled up to the drive-up window at McDonald's, where I would order a double-whatever. I did this secretly, because I didn't dare let anyone who knew me see me in

McDonalds. Once I had my order, I'd find a private place to eat. Then I'd wolf my burger down like a ravenous animal. Boy, did it taste good. I felt good afterward, too. The meat made me feel stronger and more solid on the earth. But the experience of eating meat was also very confusing. At first, I didn't know how to make sense of it. I'm eating meat, I told myself. Does this mean I'm no longer practicing macrobiotics? Unfortunately, there was more to come.

Soon, I found myself experimenting with other foods that were not on the macrobiotic diet. As my diet widened, I began to investigate even more deeply both traditional Chinese medicine and Ayurveda, the ancient healing art created thousands of years ago by Hindu sages. Ayurveda and TCM stated that all foods are considered good and potentially healing. It all depends on the person's health, physical and nutritional needs, and state of mind when he or she eats the food. There are no bad foods—only the right or wrong circumstances that such foods are eaten.

Both Ayurveda and Chinese medicine are founded on the belief that balance is the basis for health and happiness. These systems, like macrobiotics, teach that if you eat an excess of foods that make the body more contracted and tense, you will naturally crave foods that make the body more relaxed, sedated, and restful. Your cravings arise out of an unbalanced condition. In effect, cravings are leading you to a more balanced state. They are the solution, not the problem. The real problem is the underlying imbalance that is causing the cravings in the first place. These systems also teach that excesses of one or the other types of foods—either too much contraction or too much sedation—lead to unhappiness and sickness.

Out of these systems, I created a program that dramatically improved my health, brought me into a greater state of balance, and as a consequence, dramatically curtailed my cravings for foods outside the program. Once I was in a relative state of balance, and didn't have the cravings I had had before, I could satisfy my cravings either with small amounts of the desired food or with healthful substitutes. After a short time on my program, you will be able to do the same.

The First Step: Connect with Yourself Again

Among the goals of this program is to help you achieve your health goals and give you a diet that you can follow for life. But the only diet that can be sustained is the one that you create yourself. The Energy Balance Program helps you do just that. It does it by allowing you to manipulate a single aspect of the diet, which will help you shape the entire diet and make it truly your own. The part of the diet that you can alter according to your own needs is protein.

The Energy Balance Program is essentially a plant-based diet. It is low in calories, low in fat, and rich in vitamins, minerals, antioxidants, and phytochemicals—everything you need to overcome disease, restore your health, and achieve your optimal weight. But the program shows you how to manipulate your protein intake, within certain limits, to boost your energy and to promote feelings of greater mental clarity, personal power, and self-confidence. As I will show, protein intake is a very powerful tool for altering how you feel, especially if your basic diet is essentially a plant-based regimen.

I provide you with a clear and easy-to-follow program in this book. In the chapters that follow, I will describe in detail how the Energy Balance Diet can help you restore your health, achieve optimal weight, give you increased mental clarity, greater energy and immune function, and a much happier and more balanced emotional life. By the time you finish this book, I hope you will be convinced that this is the most powerful form of preventive and therapeutic medicine you have ever encountered.

But that is only one of my goals for you, perhaps even the lesser one. My higher objective is to help you experience intimacy with yourself so that you can create a way of eating and living that flows from your own inner wisdom.

My program is designed to put you in touch with your body and the innate knowing that exists inside of you. This program can strengthen your connection with your inner self. At the same time, it will reduce or eliminate the extreme foods and stimulants that create internal noise and chaos, which prevent you from knowing what you feel and what you may want.

But I think of my program as training wheels on a child's bicycle. My real intention is to help you learn how to ride the bike, which means that eventually you'll stop needing the regimented structure that I provide in this book. It is my hope that when that happens, you'll have established a working relationship with your innate wisdom so that you can pilot your life to health and happiness.

Occasionally, you may feel a need to eat foods that are outside the diet I recommend. You have my permission to do that. Whenever you do that, do it consciously. Choose the highest quality possible. Eat slowly. Chew the food thoroughly. Taste it. Enjoy it. And then experience whatever effects it has on your body. This will help your body accept the food and assimilate it more easily into your system. If you eat the food consciously, you are more likely to be satisfied with less of it. You will also know its effects on your body, which means you'll know if you made a good choice.

I believe that every so often you have to go off the diet, if for no other reason than to know the effects of different foods on your system. This is part of the process of discovering your own way of eating. I encourage you to do it, but I recommend you do it wisely.

A basic part of my program is to understand that all food has a purpose and all food can be the right choice, under the right conditions. Please do not engage in unhealthy criticism of yourself or others when you or someone you love eats something that you deem unhealthy. Recently, in one of my classes at The Institute for Integrative Nutrition, a woman started to complain to me that her son was putting purple ketchup on his food. I guess she assumed that I would join her in her criticism, but I did just the opposite. "What's wrong with purple ketchup?" I asked her. "There is nothing wrong with any food, including purple ketchup. It all depends on the person and what's right for him or her, which only he or she can determine. Obviously, purple ketchup is not the right food for you, but maybe it was balancing something in your son's condition. Maybe we should be asking ourselves which foods and behaviors were causing him to want purple ketchup in the first place. But in most cases, self-criticism, or criticism of others, creates more harm to ourselves and those we love than just eating the food we desire."

My point is to be gentle with yourself when you eat purple ketchup—or any other food that you may feel is unhealthy, but desired nonetheless. As I often tell my students: Lighten up, it's only food.

All Experience Is Food

As we all know, people eat for reasons other than just their need for nutrition. People eat as a way to love themselves, and because they are not getting enough love from others. They eat to reduce their tension, to relieve their sense of loneliness or isolation. They eat for entertainment or to feel happy. People eat because they don't get enough hugs or they're not touched enough in a loving way. They eat because they're nervous or anxious or afraid. They eat because they don't get enough exercise or because they hate their jobs. They eat because they feel disconnected from their spirit.

These needs are all forms of hunger. Unfortunately, they cannot be satisfied by food. Food can temporarily reduce the tension created by these desires, but only for a short time. Food cannot replace our need for love, or touch, or physical activity, or a rewarding job.

Love, touch, exercise, career, and spiritual fulfillment are all forms of nourishment, or what I call primary foods. Primary foods determine whether or not we are happy. The foods that we put in our mouths—the grains, vegetables, beans, and meat—are what I refer to as secondary foods. If you are getting adequate primary foods nourishment, you will find it much easier to eat healthy secondary foods. You will also discover that you need less of them in order to feel full and satisfied. The more your need for primary foods go unsatisfied, however, the greater the chances that you will abuse secondary foods.

Any dietary approach that does not take into account the effects of primary foods on your life cannot create long-term changes in your eating habits. That means you will not be able to accomplish your health goals on that program. I am going to show you how primary foods affect your eating habits, and how you can satisfy your need for primary foods so that you can eat a simple, healing diet.

So let's begin our journey to better health and inner wisdom by understanding and adopting the Energy Balance Program.

Acknowledgments

I wish to thank my parents for making everything possible. I also wish to thank Dr. Carolyn F. A. Dean, M.D., N.D., who checked the accuracy of what you'll learn in this book and provided invaluable insight and suggestions.

chapter 1

Eliminate Cravings and Lose Weight

Some time ago, a woman I'll call Jessica appeared in my office. Jessica was significantly overweight and described herself as a compulsive eater. She thought it would be impossible for her to achieve and maintain a healthy weight.

"Why would that be impossible?" I asked her.

"Because I can't stay on a diet. I'm a compulsive eater. I have overwhelming cravings and wild energy swings," she confided. "I can lose weight—I've done it many times—but I can't discipline myself for very long. For a while, I can resist the cravings, but eventually my willpower breaks and I start eating foods that I'm not supposed to eat. Once that happens, I keep eating and regain all the weight I lost, and then some."

"Why do you think your body has such strong cravings and energy swings?" I asked.

"I don't know," she said. "Maybe it's genetic. I can't seem to do the right thing when it comes to food."

"What do you feel when your cravings become intense?" I asked her.

"Tension," she said. "I feel anxious and sort of frantic. It's as if my whole body is in a state of hunger. I have to eat. I'm in such a state of gnawing, frantic anxiety that I feel out of control."

"Are you hungry for food?"

"Yes, I crave food to help me relax and feel satisfied. I'm not hungry in the sense that I'm starving. I'm hungry in a different way. I want the food to help me feel satisfied and get on an even keel."

"So," I said. "You're looking for balance."

"I guess so," she replied. "If balance means the release of tension and feeling relaxed and satisfied, then I'm looking for balance, sure."

"What foods do you eat mostly?" I asked her.

"Well, I eat a lot of protein, because I think that it helps keep my weight at least somewhat under control," she said. "I eat a lot of eggs and different kinds of meats. I try to avoid carbohydrates, because if I eat them, I'll get fatter."

"What do you eat when you give in to your craving?" I asked.

She smiled, as if I had asked a question that revealed a secret. "Usually carbs," she said, laughing. "That's what happens. I'm good while I'm on a diet, but then I start getting these cravings and pretty soon I'm eating doughnuts or muffins or pastries and stuff like that."

"What I'm going to tell you is going to sound strange at first, so let me explain," I began. "Your cravings are not your main problem. They're the solution to your problem. The real cause of your distress is your daily diet, which is creating an extreme state of imbalance. You are craving foods that will restore balance to your system. The reason the cravings are so powerful is because your diet and overall condition resulting from your diet are so imbalanced that you absolutely have to have those carbs in order to feel some degree of peace and harmony in your body."

Jessica's situation is common. Most people believe that they cannot stay on a diet because they are incapable of controlling their cravings. In fact, most diets are based on the belief that you have to control your cravings in order to succeed on the diet. Without ever saying it explicitly, the author is telling you that if you want to lose weight or regain your health, you have to control the part of you that craves foods that are not on your diet. Success depends on conforming to the diet. To do that, you have to discipline your natural instincts.

You start the program with the best of intentions. You're determined to make good on this diet. Unfortunately, you soon learn that the part of you that directs your food choices cannot be disciplined. Neither can it be controlled or suppressed or denied. Pretty soon, you find yourself eating foods that are forbidden on the diet, which is how every diet fails.

Something very interesting happens at this point. People blame themselves for failing to stick to their diets. People tell themselves that if only they could be more disciplined, they would have kept those pounds off, and finally become healthier. It never occurs to them to question the *program*. This is how people come to believe that their bodies are flawed—because they can't do what the experts tell them to do.

Let's back up a second and drop our concepts about what we're supposed to do. Let's ask ourselves a few simple questions, such as: Why is the part of your being that determines your food choices so powerful and unruly? Why can't it be controlled and disciplined? And what motivates its choices and cravings? Clearly, it's not your intellect. But what is it?

My answer is this: The part of you that cannot be controlled is actually your inner guide to health and happiness.

Your Inner Wisdom

This part of you, which is actually your innate wisdom, is always trying to make you feel better by urging you to eat foods that will dissipate, if only temporarily, your physical tension, lack of energy, and negative mood state. In essence, this part of you is always struggling to create balance, harmony, and happiness. It does so by using food as a kind of medicine to balance your current inner state.

Let me give you a few examples. People who don't sleep deeply and wake up feeling lethargic often crave coffee to boost their energy levels and clear their minds. Many who experience frustration or mild depression reach for chocolate or some other sweet food to boost their mood. After a stressful day, many want to eat something sweet or to drink an alcoholic beverage. Then, after a day of sugar and processed foods, we often feel weak and empty inside. We need something nutritious and strengthening. Thus, we find ourselves craving eggs, steak, chicken, or fish. After a thick steak, baked potato, and buttered vegetables, however, we often feel heavy, blocked, and sleepy. We need something to lighten the body and unblock digestion. So we order a sweet dessert and coffee.

Up and down we go. An intelligence that is rooted in your cells is always monitoring your physical, emotional, and psychological condition. Whenever that intelligence perceives the balance to be off—perhaps you're feeling a bit low, or tired, or constipated, or malnourished—it triggers a set of urges that are designed to restore balance and harmony to your condition. The body is always coming up with answers to resolve your current imbalance. The only way to overpower this innate wisdom is to somehow conquer your instinctual desire to be healthy, happy, and balanced, which is virtually impossible.

What we don't realize is that cravings are not the real problem, even though we blame cravings for keeping us from being able to stick to a diet. Cravings are a symptom of an underlying imbalance. By believing cravings are the problem, we fail to see where the real problem lies.

The Cravings Cycle

The real source of cravings is a vicious cycle that most Americans find themselves stuck in for all of their adult lives. The foods you use to satisfy your cravings and restore balance to your body act like a boomerang on your body, mind, and mood. Initially these foods have the desired effect. Some foods make you feel strong and powerful; others overcome stagnation and create feelings of lightness and well-being. But these mood changes are short-lived. Very soon, they wear off and the tension, anxiety, fatigue, and depression return, oftentimes worse than what they were originally. At that point, we again experience cravings for foods that provide that same short-term relief.

These same foods have terrible effects on our health, as well as on our moods and energy level. They are the basis for heart disease, cancer, adult-onset diabetes, high blood pressure, and many other serious illnesses.

As I explained all of this to Jessica, I could see that questions were popping up in her mind. "I understand that I use food to create balance and to feel relaxed and satisfied," she said, "but why does my body ask for foods that cause me to gain weight?"

"It doesn't," I said. "The intelligence in your cells only asks for substances that are sufficiently strong to restore balance. When you are only slightly out of balance, it's easier to restore balance. Let me give you an example."

Let's say that you are walking in a park and the sun is out and the temperature is perfect and the air is clean. All you want to do is bask in the sunlight and enjoy the day. Do you want food? No. Food would only spoil things, because you feel good, which is to say, you are already balanced, or close to it. You are free of all cravings at this moment.

But let's say you had a bad day at work. You come home feeling tense and upset. You're probably going to need an entire pint of Ben & Jerry's ice cream, or a couple of beers, or even a few stiff drinks to restore your sense of calm. Or let's say that you spend the day eating sugar and white rolls and drinking coffee. In the evening, you're probably going to feel weak and malnourished. Many people in that condition find themselves craving a food that they perceive will give them strength and vitality, so they go to a restaurant and order a steak with lots of mushrooms and gravy. They eat the steak and feel weighted down so they order sherbet, or tiramisu, and coffee. Once they get home, they need heartburn medication or a sleep aid. The next day they find that they are constipated and need a laxative. We are constantly responding to our imbalances by using extreme foods to restore harmony inside of us.

"What are extreme foods?" Jessica asked.

"I'm glad you asked," I said.

Extreme Foods: What Are They, and What Do They Do?

Extreme foods are foods that have an extreme effect on your body. They create strong cravings and great swings in energy. Extreme foods usually provide exactly the effect that they are eaten for. Unfortunately, extreme foods boomerang on us, meaning their positive effects are short-lived. The result is that extreme foods create exactly the opposite effects for which they are used. For example, extreme foods that we typically use to boost energy—sugar, for example—provide a short-term boost in energy, but soon result in a significant drop in energy. Extreme foods that are used to elevate mood eventually result in a decline in mood. All extreme foods cause the short-term effects for which they were used, but soon create exactly the opposite effect on the body. Once the opposite effect sets in, we find ourselves craving more extreme foods to restore energy, mood, and feelings of well-being and power.

In general, there are two groups of extreme foods, each having opposite effects on the body:

- Extreme contracting foods create short-term feelings of power and control, along with physical tension, anxiety, and cravings for extreme expanding foods.
- Extreme expanding foods create short-term elevations in mood, feelings of well-being, relaxation, satisfaction, and cravings for extreme contracting foods.

Extreme contracting and expanding foods contrast with two other types of foods, namely mildly contracting and mildly expanding. These more balanced foods create long-lasting feelings of lightness, well-being, and high energy. They also eliminate cravings.

Extreme Contracting Foods

Generally speaking, extreme contracting foods are animal foods. They provide short-term feelings of strength, nutrition, and increased physical and mental power. They are rich in protein. Among other things, the protein elevates a brain chemical called dopamine, which triggers almost immediate feelings of alertness, mental acuity, personal power, and aggression (more on brain chemistry in Chapter 3).

These characteristics make us feel more stable on the earth, more in control of ourselves and our circumstances, and better able to deal with the challenges life presents us. All of these characteristics also make us feel more compact, or contracted, and integrated within ourselves. We don't feel all over the place, so to speak, but well contained within our own limits and borders. Such feelings give us a greater sense of strength and personal security. However, when we eat extreme contracting foods in excess, we quickly feel bloated, heavy, sluggish, and slow-witted. Often, we want to sleep after such a meal, and we can almost feel ourselves gaining weight.

However, the more contracted foods we eat, the more tense our bodies become. Contracting foods make us feel blocked, especially in digestion.

They frequently result in constipation. Beef and other forms of red meat are especially difficult to digest and eliminate. Very few people, if anyone, can fully masticate a piece of steak, or any red meat for that matter. Usually, the well-chewed piece of meat goes down the esophagus as a white wad of sinew. You have no teeth in your digestive tract, so if the teeth in your mouth can't fully break down the meat, your intestines are going to have an even more difficult task. This is why meat consumption so often results in constipation, especially when it is part of a low-fiber diet.

Contracting foods cause us to crave expanding foods, which unblock and relax. The most common extreme relaxing foods are processed foods and refined white sugar.

The single substance that is contracting but contains no protein is salt, which is highly contracting. Throw salt on plants and watch them shrivel up. Salt does the same inside your body: It causes cells to become dehydrated and blood vessels and tiny organs, such as nephrons in your kidneys, to contract. The contracting effects of salt (or sodium chloride) often lead to high blood pressure, especially in salt-sensitive people.

Contracting foods create blockages throughout the body. For example, animal foods tend to be rich in fat—including saturated fat—and cholesterol. Fat and cholesterol contribute to cholesterol plaques, or blockages, in arteries throughout the body. These plaques block blood flow to the heart, brain, and other organs, causing heart attack and stroke.

The U.S. Surgeon General and other health authorities have stated that high-fat diets are associated with diminished blood flow, a dramatic increase in oxidation, the breakdown of cells, and the deformity of DNA, which can result in various types of cancer.

Animal foods are perceived as being rich in nutrients, but in fact they provide only a limited number of individual nutrients, rather than a broad spectrum of nutrition. Animal foods are typically rich in protein, iron, calcium, protein, and fat. They do not provide a broad spectrum of nutrition and are essentially devoid of antioxidants, phytochemicals, and fiber. Consequently, they often leave the body hungry for more nutrition, even craving it. Unless that hunger is satisfied with vegetables, whole grains, and fruit—all of which provide a broad spectrum of nutrients and fiber—the hunger will not be satisfied.

Here are some common extreme contracting foods, arranged from the most contracting to the least:

Salt
Beef
Pork
Ham
Hard cheese
Eggs
Chicken
Fish
Shellfish

Contracting foods cause cravings for expanding foods. The more a person eats contracting foods, the more powerful the craving will be for expanding foods.

Extreme Expanding Foods

The extreme expanding foods provide short-term feelings of lightness, elevations in mood, and relief from blockages and stagnation. The predominant nutrient that makes up extreme expanding foods is refined white sugar. There are two forms of sugar: simple, such as refined white sugar, and complex, such as those found in whole grains and vegetables. Both types are considered carbohydrates. All forms of carbohydrate, including simple sugars, cause elevations in a brain chemical called serotonin, which results in feelings of relaxation, well-being, safety, and enhanced concentration. However, refined sugar causes rapid elevations in serotonin, followed by rapid declines. When serotonin levels fall, we all typically experience feelings of depression, low-energy, anxiety, and loss of concentration.

Sugar and processed foods cause rapid and extreme elevations in both blood sugar and insulin, the hormone produced by the pancreas that allows cells to utilize sugar as fuel. Simple sugars, which are the type found most abundantly in processed foods, are rapidly absorbed by the small intestine. Once consumed, those simple sugars suddenly flood the bloodstream. The brain recognizes this immediate rise in blood sugar and signals the pancreas to secrete large amounts of insulin. As I will explain in greater detail in the next chapter, the presence of elevated insulin causes the body to burn sugar and store fat. That means that any fat contained in the meal, as well as the fat found in your tissues, is being stored when you eat a meal that includes processed foods and simple sugars. That's one of the ways people gain weight when they eat processed foods. High insulin also results in other serious disorders, including adult-onset diabetes. High insulin also causes severe cravings, especially for extreme expanding foods.

Sugar provides no nutrition; processed foods provide little or none. Sugar is a fuel for cells. That means that it stimulates metabolic activity. All metabolic activity requires nutrients, such as vitamins and minerals. Because these extreme expanding foods do not provide nutrients for metabolism, the activity that they stimulate requires the body to steal nutrients from tissues in order to conduct normal activities. Thus, these foods result in a loss of the body's vitamin and mineral supplies. The loss of nutrients causes a hunger for nutrition. Unfortunately, that hunger tends to be poorly understood. People often attempt to satisfy it with animal foods, which provide limited nutrition, or more processed foods. Consequently, the person never experiences deep satisfaction with his or her diet.

The more processed a food is, the more extreme an effect it has on the body.

Bread, Rolls, and Pastries: Expansion Foods That Cause Tension

Flour products, such as bread, rolls, and bagels, are a mix of expanding and contracting substances, which combine to create cravings for more nutritional foods. Most people satisfy such cravings with contracting foods, which result in a powerful desire for expanding foods, such as pastries and other sweets.

The cracked grain in flour products is rapidly absorbed, which means that bread and pastries cause rapid elevations in glucose and insulin, making these foods expanding. Flour products are loaded with calories and cause weight gain, another characteristic of expanding foods. But bread and rolls are often laden with salt, which is highly contracting. This combination of effects—rapid rises in glucose, insulin, and calories, plus salt—creates a food that is both expanding and contracting. The processed grain causes a rush in blood sugar and insulin, but the salt dehydrates cells and causes blood vessels and organs to contract. The combination produces tremendous tension in the body, which is why bread, rolls, bagels, muffins, and other flour products are leading causes of cravings. The more bread and rolls you eat, the more tense your body will become, and the more intense your cravings will be.

The overall effect of bread, rolls, bagels, and muffins is to cause expansion and tension. Therefore, I place them in the expanding foods category, but with the awareness that they also have contracting influences.

Milk Products: Making Small Animals into Large Ones

Included in my list is a food that is typically thought of as natural. I'm referring to cow's milk and its products, such as yogurt and ice cream. All three foods contain varying amounts of sugar, beginning with the sugar normally found in milk, lactose. Milk causes a 65-pound calf to become a 500-pound cow in less than a year. That's expansion. Like other animal products, milk is rich in protein. It also contains a limited number of other nutrients, most notably calcium. (More on milk and milk products in Chapters 3 and 4.)

Expanding foods cause cravings for contracting foods. The more a person eats expanding foods, the more extreme the craving will be for contracting foods.

Here are some common examples of extreme expanding foods. They are arranged in order from most expanding to least expanding:

White sugar
Pastries
Doughnuts
White rolls
Bagels
White bread
Whole-grain bread
Foods that contain artificial ingredients
Cow's milk
Coffee and other caffeinated beverages
Soft drinks

Jell-O and puddings
Candy
Chocolate
Alcoholic beverages

The following table summarizes the respective effects of contracting and expanding foods.

Characteristics and Effects of Extreme Expanding and Contracting Foods

Extreme Contracting Foods	Extreme Expanding Foods
Dramatically increase brain levels of dopamine.	Dramatically increase brain levels of serotonin.
Create feelings of vitality, personal power, and strength.	Create feelings of relaxation, enhanced digestion, and elevations in mood and well-being.
Tend to be rich in small groups of nutrients, such as protein, calcium, iron, and vitamin B12.	Highly processed.
Deficient in antioxidants, phytochemicals, and a wide spectrum of vitamins and minerals.	Low or devoid of nutrition.
Rich in fat and protein.	Rich in calories.
Create contraction, blockages, and stagnation.	Cause extreme and rapid increases in blood sugar and insulin.
Cause weight retention and weight gain.	Cause weight retention and weight gain.
Usually difficult to digest as they lack fiber and are highly constipating.	Usually difficult to digest as they lack fiber and are highly constipating.
Contribute to heart disease, cancer, digestive disorders, and other serious illnesses.	Expansive foods often have the effect of unblocking stagnation, especially in the digestive tract.
	Expansive foods result in a net loss of nutrients in the body.
Contracting foods create strong to intense cravings for expanding foods.	Create strong to intense cravings for contracting foods.

As you can see, the standard American diet is dominated by foods that have an extreme effect on the body and the mind. Indeed, extreme foods do more than just toss us around from craving to craving. They create an inner life that is characterized by so much tumult, chaos, and confusion. We are being tossed around by the extreme energies of such foods. Extreme foods create extreme mood swings—great highs and nasty lows, anger, and sadness. We're tossed from one extreme emotion to the next. It's like living within a storm.

Balanced Foods

Like extreme foods, balanced foods have contracting and expanding influences, only their contracting and expanding effects are more mild and gentle. In general, whole grains, such as brown rice, barley, millet, corn, oats, buckwheat, and whole wheat, are mildly contracting.

Vegetables, including root, round, stalk, leafy, and green vegetables are all mildly expanding, with roots being the least expanding and leafy vegetables being the most expanding, though still only mildly so. Fruit is more expanding than vegetables, and fruit juice is even more expanding than fruit.

Balanced foods are rich in nutrients, which means they are deeply nourishing. They don't cause cravings for more nutrition. The nutrients in balanced foods are easy to digest and assimilate. Balanced foods do not cause constipation, or indigestion, or heartburn. Just the opposite, they promote healthy and effortless digestion. You don't need drugs to restore bowel elimination or correct indigestion or heartburn, all common consequences of extreme foods. Balanced foods create long, gentle rises and falls in blood sugar, not the extreme rises and falls in blood sugar and insulin normally associated with extreme foods. Balanced foods promote optimal circulation of oxygen and nutrients to cells. They are low in fat and cholesterol. Balanced foods are low in calories and consequently promote optimal weight. They are associated with greater health and longevity.

The most balanced foods in the food supply are plant foods. They are rich in complex carbohydrates, which means that they provide abundant and long-lasting energy without creating dramatic elevations in blood sugar or insulin. They are abundant sources of vitamins, minerals, antioxidants, and phytochemicals. They're rich in fiber, which makes them easy to digest. Plant foods are very easy on the body. I often think of them as virtually invisible to my body. They just blend in, without requiring much effort or energy from my body. They have virtually no negative side effects.

Balanced Foods

Mildly Contracting Foods	Mildly Expanding Foods
Whole grains	Vegetables and fruit
Amaranth	Root vegetables, such as carrots
Barley Brown rice	Round vegetables, such as squash, pumpkin, and onions
Buckwheat	Green and leafy vegetables
Millet	Fruit
Oats	Fruit juice
Quinoa	
Teff	
Noodles or pasta	

Unlike extreme foods, balanced foods create milder and less demanding cravings. These cravings are more easily satisfied, either with healthful substitutions or with smaller amounts of the desired extreme food. So if you eat a diet that is dominated by balanced foods and you find yourself craving chocolate, or a beer, or a glass of wine, you'll be satisfied with a much smaller amount than if your diet was dominated by extreme foods.

Balanced foods permit you to be much more in control of what you are eating. They give you greater freedom in your food choices. They also have a very different, and profoundly positive, effect on your health and weight. Balanced foods promote weight loss, without the hunger, cravings, and feelings of being out of control.

When you are in a balanced state, you are more likely to make choices that sustain balance and support your health. The opposite is also true: when you are extremely out of balance, you are more likely to make food and behavioral choices that have extreme effects on your body, mind, and environment. Those choices are likely to sustain your extreme imbalanced condition.

The Energy Balance Program is made up primarily of balanced foods, which are supplemented with smaller amounts of extreme foods.

The Energy Balance Program

The Energy Balance Program is based on four food groups, each of which provides its own array of healing properties for controlling energy levels, reducing weight, and restoring health. In Chapter 4, I show you a step-by-step approach to adopting the program. In addition to the four food groups listed later are many condiments, snacks, and desserts. (I have also provided menu plans and more than 100 recipes in Chapter 10.)

Group 1: Green and Leafy Vegetables

These include broccoli, cabbage, collard greens, kale, mustard greens, and watercress. These foods, and others like them, provide an abundance of vitamins, minerals, phytochemicals, and immune-boosting and cancer-fighting substances that restore health. As I demonstrate in the following sections, these foods are also extremely low in calories, which means they promote rapid weight loss. I recommend that you eat at least three servings from any of the foods from this group per day.

Group 2: Sweet Vegetables

These include parsnips, the many varieties of squash, carrots, onions, sweet potatoes, and yams. These vegetables provide energy, flavor, and satiety. They also provide gentle elevations in blood sugar, without causing the extreme highs or lows associated with processed foods. I recommend that you eat at least one serving of the foods from this group per day.

Group 3: Protein Foods

These include red meat, chicken, fish, eggs, and beans. During the first month on the diet, the protein foods can come from animal or vegetable sources. If the protein source is from animal foods, such as red meat, chicken, or fish, I encourage you to eat only 3½ ounces, the size of a deck of cards. If the protein food is eggs, I recommend two to four eggs per week.

After one month on the program—or sooner, if you prefer—I recommend that you reduce your animal protein consumption to one time per day. The other two protein foods should come from plant sources, such as beans or bean products, including as tofu, tempeh, or some other soy product. You are still permitted to eat up to one serving of animal foods per day.

Group 4: Whole, Unprocessed Grains

Whole grains are grains that remain largely in their original state, meaning as nature created them. In the vast majority of cases, they have not been processed or altered. Whole grains include brown rice, barley, buckwheat, corn, millet, and wheat. Whole grains are absorbed slowly by the body, thus keeping glucose and insulin levels low. At the same time, they provide long-lasting energy. This combination gives grain the unique ability to prevent the steep decline in blood sugar levels that gives rise to hypoglycemia and its characteristic symptoms, such as fatigue, mood swings, food cravings, and other physical and emotional consequences.

I encourage people to eat specific foods between meals and to adopt certain behaviors that promote health and high energy. (The program is fully explained in Chapter 4.)

Jessica's Success

I placed Jessica on my Energy Balance Program. In less than a month, Jessica was eating a plant-based diet that was supplemented with animal foods four times per week. "I couldn't believe how much my cravings diminished," she said. "It felt as if I had my life back again." Within 6 months of starting the program, she had lost 30 pounds, and within a year, she had lost 50. "I still have a little ways to go," Jessica said recently, "but I have come a long way and I'm way better than I've ever been. But best of all, I feel great. I'm not being tossed around by my food cravings and the crazy mood swings that I used to suffer constantly."

Balance Restores Your Connection with Your Inner Wisdom

As most people already realize, food has a profound effect on our mental and emotional conditions. We all use food to improve how we feel. Unfortunately, the extreme foods that we typically use to improve our moods and energy levels backfire on us. They cause extreme swings of energy—both mental and physical—like a wave. And trying to control those peaks and valleys with food causes short-lived results. An increase in energy is quickly followed by a decline, resulting in an increase in tension, an elevation in the intensity of our cravings, and even more profound mood swings, including depression.

Balanced foods have a very different effect on us. They bring a much greater experience of peace and equilibrium, and a deeper sense of connection with your inner self. When your diet is dominated by balancing foods, your overall condition, including your mental and emotional states, become more balanced and under control.

Once we achieve a more balanced state, changes begin to occur within us that we do not originally anticipate.

First, our cravings for extreme foods and our extreme behaviors gradually begin to diminish and our ability to maintain a healthful way of eating becomes easier and easier. This, of course, has a profound effect on our health. As we sustain a diet that is composed chiefly of balanced foods, the body is given what it needs to heal itself. On a daily basis, the body is fed an abundance of vitamins, minerals, antioxidants, phytochemicals, and fiber, all of which strengthen the immune system and promote the body's healing mechanisms.

Meanwhile, balanced foods are low in calories, which means that they promote weight loss. As I will show in Chapter 3, balanced foods speed weight loss even further by keeping insulin levels low, which promotes constant fat burning. Under such conditions, the body becomes stronger and healthier.

Free of the constant tumult created by extreme foods, the mind works with greater clarity and insight. And like a great sea that has been tossed by storms for years and years, the mind becomes increasingly calm and tranquil.

In these ways, balanced foods have a profound effect on our inner life.

Coming into Balance

As the inner world becomes more balanced and peaceful, something unexpected begins to happen. Gradually, you become aware that your choices are leading to greater and greater happiness. Those choices also tend to sustain and promote your health. The reason this is happening is because you have greater access to your innate wisdom. This is not as mysterious or mystical as it might seem. In fact, you have experienced such moments of clarity many times already.

I'm sure you've had the experience of walking into your house, expecting someone to be there, and discovering that no one is home. You're alone.

Whatever your initial reaction may be, your brain waves adapt to the situation and you soon find yourself entering a deeper and more peaceful relationship with yourself. It's quiet. There's no one making demands of you. No one is pulling at you for attention. Your breathing gets deeper and slower. You relax and settle into a more intimate connection with yourself. You realize that you are free to do anything you want. If you like, you can just sit on the couch and stroke the cat. In that state of peace and tranquillity, you have access to the deeper levels of your true nature and innate wisdom. If you ask yourself what you would like to eat, I am betting that you would make a delicious, healthful, and satisfying food choice. Let's say that you have certain health goals; perhaps you want to lose weight. From such a state of balance, your choice not only would satisfy your palate, but would promote weight loss, as well.

But let's say that you face a difficult decision. If you asked yourself in that same state of balance and connection with your inner self what you should do, the chances are that you would feel your way to the answer. The answer would arise from your cells, from your deep and inner knowing. Not only would you know what you wanted to do, you would see how the situation was affecting you, how it needed to change, and why you must do this or that to make the circumstances right for you.

The inner intelligence that we all possess does not concern only our food choices, but is there to guide us in every decision. The trouble most of us face is that because we eat extreme foods, and consequently live more extreme lives, we cannot access that innate wisdom.

Balance is the key to health, emotional and psychological happiness, and clear perception of the path that leads to our greatest joy. Achieving a balanced state should be our top priority when it comes to our food choices. Everything that you desire—optimal weight, better health, self-confidence, and clarity of purpose—flows from such choices.

Sally's Story

Sally was about 25 pounds overweight when she adopted the Energy Balance Program. Although she was a vegetarian, she ate lots of dairy products and processed foods.

"When I first started the program, I thought that I had been eating pretty well," Sally recalled. "I was also very dogmatic about being a vegetarian. But one of the first things I learned on the program was that not everyone should be a vegetarian. People have different nutritional needs. Some people really need to eat meat in order to feel strong and centered.

"While I still maintained my vegetarian lifestyle, I stopped eating dairy foods, bread, salty snacks, and processed foods. I didn't realize it until I changed my diet, but bread and salty snacks were creating so much tension in my body. I was constantly craving food just to relieve my body of tension.

"Following the Energy Balance food groups, I lost weight without even trying. I also felt a lot stronger, more centered, more confident, and a lot more flexible. And understanding why some people need meat made me much more understanding of my partner at the time. I wasn't being critical about the fact that he was eating meat. That changed our relationship. I no longer felt I had to change my partner, and that made our relationship a lot better." Sally ended up losing 20 pounds. Her skin became brighter and more youthful, her body more relaxed and flexible. She also became more energetic and more confident. People don't realize the subtle effects that food can have on the body until they start to eat in a more balanced way.

In Chapter 4, I offer a program that can give you a step-by-step approach to dietary change and improving your health. But just as an experiment, eliminate processed foods for one week, especially bread and salty snacks, and then see what effects this simple step can have on your weight, feelings of anxiety and tension, and food cravings. You will notice significant changes in just one week and you'll begin to see how easy it is to lose weight, overcome cravings for foods that you know are causing you problems, and achieve better health.

chapter 2

The Energy Wave: Secret to Your Ideal Weight

Last year, a mother brought her 11-year-old son, Kevin, to me and asked how she could help him deal with his weight and eating problem. Her son was significantly overweight, and she didn't know what to do for him.

"Kevin is constantly hungry," she said. "And he's constantly eating! Everywhere he goes, he wants a soda or a snack. He can't go an hour without food. I keep wondering, how can he be this hungry?!"

"Why don't you tell me what he eats during the day?" I asked the mother. She then proceeded to list all the foods she could think of that her son routinely ate. Everything she mentioned was processed, from the processed sugar cereal in the morning, to the peanut-butter-and-jelly sandwiches on white bread for lunch, to the pizza, microwave meals, and the regular stops at fast-food outlets for dinner. In between the meals was an endless supply of processed snacks and desserts, everything from cheese crackers and chips, to soda pop and cookies.

"Maybe he's not hungry for calories," I said. "Maybe he's hungry for nutrition."

"What do you mean?" she asked. "I'm feeding him all day long. I would think that he's getting too much nutrition."

"The food he's eating is rich in simple sugars but deficient in nutrition. Sugar is a fuel for cells. It makes cells work, but they need vitamins and minerals to do their jobs. So he's fueling his body, making cells work, but not giving them the raw materials they need. The only thing the body can do is crave more food so that it can get the nutrients it needs. He's on a very inefficient diet—he needs to eat a lot of food to get just enough of the nutrients to operate his body. That kind of imbalance keeps him craving more food."

"What can I do?" the mother said, now a bit stunned.

"He's got to reverse the formula," I said. "Eat foods that are rich in nutrients and low in calories—the exact opposite of what he's getting now."

I then laid out my program for healing her son—including exercise. Six months later, he had lost 30 pounds. Needless to say, Kevin and his mother were thrilled.

Our craving for sugar is as natural as our desire for air. During our two million years of evolution, nature genetically programmed humans to desire sweet-tasting foods over all others. The reason: Unprocessed sweet foods are the greatest sources of energy and nutrition.

Long before there was food processing, the only source of sweet taste were plant foods, such as squash, tubers, roots, grains, and fruit. In order to get the sweet taste that the body desired, people had to eat plants. It was no coincidence that these same sweet foods were also the greatest sources of nutrients, energy, and fiber—everything we needed to maintain health and survival. (We don't call it "Mother Nature" for nothing.)

Food processing didn't really begin until civilization arose, some 10,000 to 14,000 years ago. At that point, people began making whole-grain flour, sugar, beer, wine, pickles, and many other fermented foods. We also harvested honey. But these foods were a small part of the overall diet. Most of the foods we ate were whole and unprocessed vegetables, grains, beans, and fruit. It didn't matter that people ate small amounts of sugar and honey and drank some wine or beer. Those foods were for special occasions; they weren't everyday fare.

So for about two million years, people ate food essentially as nature produced it. Yes, we also ate animal foods, but in much smaller amounts than plants, primarily because the plant foods were so much more abundant, and animal foods were more difficult to come by.

Then along came the middle of the 1900s and food processing exploded. Suddenly, sugar was in everything, from catsup to toothpaste. That wasn't enough, however. We turned whole-grain bread into white bread, brown rolls into white rolls. We did it by stripping the grain of its fiber and the nutrient-rich germ.

Pastries, muffins, bagels, and doughnuts—every grain product we ate was straight from the food-processing plant. The most abundant foods in supermarkets today are processed foods, everything from soft drinks, snacks, already-prepared meals, desserts, and condiments. Nearly everything contains artificial sweeteners, colors, flavors, and preservatives.

People of the nineteenth and early twentieth centuries would not recognize food in today's supermarket, especially the foods generally eaten by children, which typically are the most artificial and highly processed. Children eat the most sugar, artificial ingredients, and highly refined foods available, and it shows. Obesity is epidemic among children today.

All of this has happened because we love sweet taste, and we managed to create a lot of it by extracting the sugars from the foods in which they originated. In the process, we transformed our natural desire for sweet taste into a ravenous, insatiable monster within us. And that monster has been leading us in the wrong directions ever since.

Decreasing Nutrition, Increasing Calories

Most people today don't realize the extreme effects processed foods are having on us. Processing strips the food of nutrients. Some are reintroduced during "fortification," but there is no way that a laboratory can reintroduce all the vitamins, minerals, carotenoids, phytochemicals, and fiber that were originally in that food. Did you know that a single tomato contains more than 10,000 phytochemicals? How do you put all of that back into a food after you've taken it out? You can't do it.

The recent research on antioxidants, carotenoids, and other phytochemicals has started to awaken us to how devastating the loss of plant chemicals has been to our health. The vast majority of illnesses from which we suffer today, including heart disease, cancer, Alzheimer's, Parkinson's, arthritis, asthma, cataracts, and glaucoma, are all caused by a process called oxidation. Oxidation essentially is the underlying cause of all aging and occurs throughout nature. It's the same process that causes iron to rust and an apple to brown and then spoil. Oxidation causes cells to die or become deformed. In many cases, normal tissues are transformed into scar tissues. When that happens in the tissues of your skin, you start to wrinkle and age. When it happens in your muscles, or liver, or kidneys, these tissues and organs lose their functional capacity. You literally begin to shrivel up. In many cases, oxidants interact with the cell's DNA, causing it to mutate and become cancerous. The antidote to oxidants is *anti*oxidants, which slow and sometimes stop oxidation from occurring. In this way, antioxidants slow the aging process and protect us from disease. Not only do antioxidants stop oxidation, but they also boost our immune and cancer-fighting defenses.

Carotenoids are the chemicals in plants that create color. Many of them serve as antioxidants. Others act as immune boosters and cancer fighters. Phytochemicals simply mean plant chemicals. As I mentioned earlier, there are thousands of them. Like carotenoids, many act as antioxidants, while others boost immunity and protect us from all kinds of illnesses, including cancer. Humans had been eating whole, unprocessed foods for two million years. And then in the space of a single generation, we stripped the food of its nutrients to make it sweeter. But in the process, we created a food supply that had a tiny fraction of the nutrients and phytochemicals that were in the diet we were used to eating. Yes, the food is sweeter today, but it's a whole lot less nutritious. This leaves the body wanting for nutrients, so naturally we have cravings. Nature programmed us to eat food that was bursting with nutrients, antioxidants, phytochemicals, and fiber—foods that contained myriad and mysterious chemicals and energies. And we turned it into cardboard—or something close to it.

No one can be fully satisfied on a diet that is processed and artificial.

The loss of nutrition and plant chemicals is one of the reasons why people crave food, even when they're full and they're obese. Our need for nutrition is not the only source of cravings, however. The biggest creator of cravings is the type of carbohydrates we eat today.

Complex vs. Simple Carbs

Carbohydrate is a dirty word today. But you have to remember that we're not eating the types of carbohydrates that nature intended for us to eat. We're eating something that's been deformed, denatured, and devaluated.

The type of carbs that appear in nature—that is, as part of whole foods—are "complex," meaning they are composed of long chains of sugars. These long chains are bound up within the food's fiber. The sugars inside complex carbs have to be broken free from both their chains and the fiber if they are to be released into your bloodstream. The only way that can happen is if they are thoroughly chewed and then acted upon inside your small intestine.

Absorption of these sugars is relatively slow and methodical. The sugars are broken free and taken into the bloodstream over the course of many hours, which is why complex carbohydrates provide long-lasting energy. If you eat a bowl of cooked grain in the morning, you will very likely have energy throughout the entire morning, before you experience a dip in energy around noon—just in time for you to consume another round of complex carbohydrates to take you to dinner.

An interesting characteristic of the complex carbs in plant foods is that their sweet flavor is released only after they have been chewed thoroughly. Complex carbohydrates start to be broken down in the mouth by an enzyme in saliva known as amylase. Only by chewing the food thoroughly and mixing it with amylase can some of the carbohydrates be released so that their sweetness can be tasted. Sweetness is a reward for chewing. Chewing makes the food more enjoyable. And of course, it enhances digestion and assimilation of nutrients.

A very different process occurs when you eat processed foods. The sugars in processed foods are simple, meaning they are already free of their chains and fiber. The sweetness is tasted immediately. Also, the simple sugars start entering your bloodstream through your tongue the moment you start eating them.

Processed foods don't need to be chewed and they don't need to be digested in your intestines in order to send their sugars into your bloodstream. There's a law in the body that goes like this: If you don't use it, you lose it. For example, if you don't exercise your muscles, they will wither away. It's the same with your bones, eyes, and brain. In the same way, if you don't eat fibrous foods that work your intestinal tract, you will weaken your digestive system. The number-one form of over-the-counter medication sold today is digestive aids—everything from antacids and drugs for diarrhea, to laxatives and substances to counteract bloating.

Clearly, we've got a problem with our intestines. The reason: We don't eat fiber-rich plant foods that promote healthy digestion.

Processed Foods Are Loaded with Calories

Sugar is a form of energy or fuel. This means that it provides calories. A calorie is a unit that measures the amount of energy in a given food. The more sugar in the food, the more energy, or calories. The more calories in a food, the more likely it is to cause weight gain.

Processed foods contain lots of sugars and therefore lots of calories. In fact, processing concentrates the calories, meaning it packs more calories into a smaller volume of food, as Robert Pritikin points out in his book *The Pritikin Principle: The Calorie Density Solution* (Time Life Books, 2000). Pritikin's book, which provides the caloric density for hundreds of foods, reveals just how rich in calories processed foods really are. For example, white bread, which is highly processed, contains 1,210 calories per pound. Broccoli, which is not processed, contains only 130 calories per pound. One of the big reasons why broccoli contains so few calories is because it is loaded with water and fiber, neither of which contain calories. A pound of pretzels, another highly processed food, provides 1,770 calories, while a pound of apples contains only 270 calories. Apples contain lots of water and fiber. A pound of Oreo cookies contain 2,200 calories, while a pound of strawberries only 140.

In general, processed foods are loaded with calories, while whole and unprocessed foods contain very few calories. If you want to lose weight fast, you don't have to go on a high-protein diet. Just eat a diet that is 90 percent vegetables for a while, and you'll lose weight real fast. Not coincidentally, your health will improve just as rapidly.

As I explained in Chapter 1, I regard processed foods as extreme foods, because they have extreme effects on the body. One of those extreme effects is how processed foods affect your insulin levels, which in turn make you fat.

Processed Food and High Insulin

When refined white sugar enters your bloodstream, it becomes glucose, or blood sugar. In response to the presence of sugar, your pancreas secretes insulin, the hormone that allows glucose to pass inside the cell membrane and be utilized as fuel. Insulin acts as the cell's gatekeeper for blood sugar or glucose.

The sugars in processed foods are rapidly absorbed. Not only that, there are a lot of them in processed foods. When you eat an Oreo cookie, or a candy bar, or even white bread, your bloodstream is suddenly flooded with sugars. Your pancreas reacts by secreting lots of insulin. Now you've got high blood sugar and high insulin. Your body sees this as an emergency situation that could trigger a serious threat to health. It's got to do something with all that sugar and insulin. One of its options is to burn it.

Your body, in its wisdom, says, "I've got to burn this extra sugar as quickly as I can. So instead of burning both sugar and fat, the body says, I'm going to burn only sugar. That will bring down the sugar levels as quickly as possible."

That's a smart move on the body's part, but, unfortunately, the natural result is weight gain. Let me explain why.

Your normal fuel mix is actually a combination of sugar and fat. Even now, as you read this book, you're probably burning both fat and sugar at the same time—usually about 50 percent sugar and 50 percent fat. But when glucose and insulin levels spike, the body decides to burn only sugar to protect itself against excess blood sugar levels. Instead of burning fat, the body stores all its fat—that means the fat that's in your bloodstream, the fat that's in your meal, and the fat that's in your tissues. So let's say that you just ate a cheese Danish, which contains lots of sugar and fat. Your body will burn the sugar fast and store the fat circulating in your blood, the fat in the cheese Danish, and the fat in your tissues. In addition, whatever sugars the body cannot burn as fuel, it will store as fat. Remember, processed foods are loaded with calories. As not all of them can be burned, many are stored as fat, too.

That's how processed foods cause weight gain.

What most people do in America is eat a high-fat meal, such as a double-whatever from a fast-food outlet, with French fries, a milkshake, or a soda. Maybe they also add a dessert afterward. There's enough sugar in the soda or milkshake to force the body to stop burning fat. There's also a lot of fat in the French fries, milkshake, and double-whatever. Now the body is loaded with sugar and fat. And it's got to burn the sugar fast. So it burns the sugar and stores the fat. Any calories that are left over will be stored, too. Stored fat and calories mean fatter people.

Here's the formula for overweight and obesity:

High glucose + high insulin + high calories + high fat = guaranteed weight gain

The standard Western diet provides this formula at every meal. No wonder we suffer from rampant obesity!

Whole Foods Keep Insulin Levels Low

A diet made up largely of whole grains, vegetables, beans, fruit, and smaller amounts of animal products does just the opposite: It keeps glucose and insulin levels down, which results in continually burning both sugar and fat. Also, there are relatively fewer calories in meals composed of these foods.

Unprocessed plant foods provide a slow infusion of sugars into the bloodstream, which means insulin levels and glucose levels remain low. Unprocessed plant foods are also low in fat.

Some people argue that some vegetable foods are rapidly absorbed. It's true, vegetables like carrots and parsnips are, but those foods are so low in calories that it doesn't matter. There are hardly any sugars, or calories, in those foods to promote weight gain.

Put a person on a vegetable diet and watch that person lose weight fast. Why? Because plant foods are so low in calories that they force the body to burn its own fat. Nobody gets fat on a diet that's made up largely of green vegetables, sweet vegetables, whole grains, and some low-fat animal foods. But throw in a bunch of Oreo cookies, a lot of bread, french fries, high-fat salad dressing, and a few quarter-pounders, and you've got yourself a weight problem.

Here's a formula for weight loss that will work:

Low glucose levels + low insulin + low calories + low fat = substantial weight loss

Under these conditions, you will burn fat even when you're sitting down watching television. If you exercise regularly, you'll lose weight even faster. The best way to keep glucose, insulin, and fat levels low is by eating an abundance of whole, unprocessed plant foods.

The human body is a unified bio-computer. If one part of our lives is out of balance, the whole system gets out of balance. The same formula that makes people gain weight also causes emotional distress and severe mood swings. Let me show you how.

Cravings and Mood

At my school in New York, the Institute for Integrative Nutrition, I constantly try to encourage my students with positive messages about themselves. I am training them to counsel people in matters of health, diet, and lifestyle. But in order to get them out into the world and doing that work, I first have to help them believe in themselves. Right at the beginning of a new semester, one of my students, a charming young woman who appeared bubbly and happy, came to me and said that she was miserable and irritable all the time.

"All the time?" I asked her. "You seem happy in class."

"Well, not all the time," she admitted. "But I easily get irritable. I can't help it. I get to work with all the right intentions and by 10:30 or 11 in the morning I feel rotten, and weak, and on edge. Maybe I should get some therapy, because I feel like I'm a bad person or something. I'm starting to have doubts that I can work with people."

"Well, there's nothing wrong with therapy," I said. "But many things that people say are psychological, I say are physiological."

"What do you mean?" she asked me.

"Sometimes the condition of the body is the real source of the imbalance," I said. "It's not our parents or something that happened in the past. It's the way we are taking care of ourselves right now. If we abuse the body, it doesn't function as well, and that can affect our energy levels and moods. How do you feel when you first get to work?"

"Good," she said. "I want to do a good job. I'm raring to go. I like people and I like myself. But all of that changes later on."

"Do you sleep well?" I asked. "How do you feel when you first get up?"

"I sleep well, it seems, and I feel pretty good when I wake up."

"So you feel good in the morning until about 10:30, and then you turn into Godzilla. Right?"

"Yea, by 10:30 or so, I'm a monster. It happens in the afternoon, too."

"And what do you do when you start to feel bad?" I asked.

"Usually I have a cup of coffee or a snack. That picks me up a little bit, but sometimes the coffee makes me anxious and jittery, which only makes me more moody and irritable."

"What do you have for breakfast?" I asked.

"I like sweet foods in the morning. I know I'm not supposed to eat this, but I have a bagel or sometimes a Krispy Kreme doughnut. Sometimes I'm good, though. I eat some fruit or a granola bar. I also have a cup of coffee."

"And what do you have for lunch?" I asked.

"I'm good at lunch. I usually have a salad or a sandwich on seven-grain bread."

"Okay," I said. "Let's start with breakfast. I want you to do an experiment. I'm going to give you five different breakfasts and I want you to eat each one over five days. I want you to pay careful attention to how you feel at 10:30. See if you feel different after eating each of these breakfasts."

I then listed the five breakfasts:

- On Monday, I asked her to eat a big bowl of oatmeal with raisins and to drink a cup of hot tea. No coffee.
- On Tuesday, eat a bowl of brown rice and sweeten it with rice syrup. Tea again.
- On Wednesday, eat two eggs and some toast, without butter. No coffee, but black tea is fine.
- On Thursday, eat a variety of fruit, such as cantaloupe, apple, and pear slices. Maybe add some blueberries or strawberries. Tea.
- On Friday, eat some smoked salmon on a piece of whole-grain toast, with lettuce and other vegetables. Drink some tea with it.

"I want you to keep a journal all week long and at 10:30 or 11 A.M., write down what you feel each morning. Make a short but detailed note. See if there are any differences, depending on the breakfast you ate. Next week, come back and tell me how you felt after each breakfast."

As it turned out, my young friend had a very different energetic and emotional experience that week. The first thing she noticed on Monday was that her mood did not dip at 10:30. In fact, she noted that she had good energy and felt fairly stable emotionally until lunchtime, when she started to feel weak and hungry. The same thing occurred on Tuesday when she had brown rice and rice syrup for breakfast. On Wednesday, the eggs made her feel very aggressive, tight, and strong, but she became constipated and felt it was from the eggs, which she didn't often eat for breakfast. On Thursday, the morning she ate fruit, she felt weak and irritable. "I felt as if I lacked any power inside," she told me. On Friday, she felt stronger and more in charge of herself.

When she explained all of this, I asked her if she still thought she needed therapy in order to deal with her problem.

"No," she said with a bright smile. "I think it's the food that's causing the problem."

"Yes," I replied. "It's the food. Many people feel great when they eat fruit for breakfast," I said. "Others like protein in the morning, such as eggs. Each of us has to discover what our bodies need, and what makes us feel good, in the morning, at lunch, and at dinner."

This woman experienced greater emotional stability when she ate whole grains or protein, such as eggs and fish. But the eggs affected her digestion adversely, so the grain is better for her. What was clear to me, however, was that a processed food at breakfast was ruining her morning and making her believe that there was something wrong with her. Many people fall into this same trap.

Understanding the Energy Wave

Most of us attribute changes in our energy or mood to sleep deprivation or excess work or other factors in our lives that have little or nothing to do with our diets. Moreover, even when we note that these patterns have some consistency—a dip in energy at 10:30 A.M., perhaps, and another in both energy and mood at 3 or 4 P.M., and still another around 7 P.M.—we rarely see such fluctuations as directly linked to our way of eating. Yet our diets are directly responsible for them.

Processed foods and refined sugars drive energy and mood levels upward, at least temporarily. Once blood sugar levels ebb, as they do within an hour or two after eating a processed food, you find yourself in a hypoglycemic state, with many of its associated symptoms. Among the most common symptoms are anxiety, fear, mood swings, depression, shaking of limbs and fingers, and significant fatigue. Many get cold hands and feet. Others feel a distinct loss of will.

This rise and fall is a distinct pattern, or energy wave, that not only determines your energy levels, but also shapes personality and behavior. When you find yourself in a hypoglycemic state, or what might be called the trough of the blood sugar wave, cravings arise for the types of foods that will catapult you out of the trough, or hypoglycemic state. Of course, at that moment, most of us eat a food that will reproduce that same extreme blood sugar wave, which means that it's only a matter of time before we find ourselves in the trough again, suffering another round of hypoglycemic symptoms and cravings.

Because we are creatures of habit, this wave pattern eventually becomes a way of life, an unconscious pattern. Indeed, our food dependencies become our "little pleasures," the ways in which we reward ourselves or soothe away our pains. Food becomes a way in which we give ourselves love.

While many people use this term—loving themselves with food—very few of us stop to consider what we mean. In essence, we are saying that we are using food to release ourselves from stress—or mental, emotional, and physical tension. We are using food to temporarily reduce or eliminate psychological distress and/or physical tension. The tension-relieving effects of that food wear off, which is when cravings for that food arise again. This, at its very core, is a kind of food dependency or addiction.

Helping us to keep our habituated patterns secret, even from ourselves, is the fact that the foods we eat habitually are widely accepted and available. Thus, we needn't think twice about drinking another cup of coffee; or eating a pastry or doughnut at 10 or 11 A.M.; or a chocolate bar or bag of chips at 2 or 3 P.M.; or another drink at 6 or 7 in the evening. These foods are the tools we use to manipulate our energy wave and to restore feelings of well-being, optimism, and relaxation.

Unfortunately, many of the foods we use to manipulate our mood and energy are high in calories and toxic substances that cause weight gain and illness. They also fail to maintain our mood or energy. Indeed, after causing an initial state of relaxation, their positive effects quickly wear off, leaving us feeling weak, irritable, and physically tense all over again. Eventually, these low states, with periodic and short-lived highs, become our normal state, one that we take for granted, without ever linking it to our diet.

Food manufacturers have provided an array of processed foods with exceedingly high doses of sugar, caffeine, salt, fat, and other chemical substances that are the source of our dependency. Many of the most extreme foods are marketed specifically for children, which is when this food-dependence begins for most people today. Once it takes hold, the addiction continues for most people through the rest of their lives. In time, we grow more and more dependent on extreme foods, especially those that contain refined sugar, processed grains, and caffeine, in order to manipulate our systems each day.

Eventually, the rise and fall of our energy wave becomes a "normal" pattern, or one that we hardly notice, and most certainly do not link to our weight, appearance, or quality of life. That is, until we try to change it. Once we try to give up the foods to which we are addicted, we suffer withdrawal symptoms. In fact, it isn't just withdrawal from certain foods, but a deviation from an energy wave pattern that we have become accustomed to and that supports the lifestyle we have grown accustomed to.

All addictions, no matter whether they are caused by drugs or food, can be seen as a wave pattern in which a specific substance causes a temporary rise in energy, mood, and outlook. The rise in energy and mood endures for a certain period of time and then falls. When they fall, many biological and psychological factors experience a dip as well. At that point, cravings for the addictive substance are stimulated.

This wave pattern, which I refer to as the energy wave, alters our feelings, thinking, and behavior. In a very real and fundamental way, it drives our lives. In fact, long before people become addicted to alcohol or drugs, the patterns of addiction are already present with food. Alcohol and drugs are just more extreme substances, having the same, but more powerful, effects as food.

The key to understanding addiction is to recognize that it is an extreme energy wave in which there is a rapid elevation in energy, mood, and outlook, followed by a dramatic decline in all three, and that this wave is driven by a substance—in this case, food. In fact, this steep rise and fall, or extreme wave, is the basis for all forms of addiction, including those involving food.

The most common outward signs of our dependence are overweight, fluctuating energy levels, chronic fatigue, hypoglycemia, mood swings, and alternating periods of anxiety and depression. Eventually, our dependency upon highly toxic foods results in severe illness. However, long before our fates are sealed, these substances control our food preferences and the quality of our lives.

Get Off the Roller Coaster

The key to healing this cycle of addiction is to extend and moderate the energy wave. Rather than have an energy cycle that rises quickly and then falls precipitously, we must have an energy wave that rises and remains up for several hours. When it starts to dip, it must dip slowly and gently, rather than taking a steep plunge. In other words, the highs must endure and the lows must not be as extreme.

We do that by eating slowly absorbed, complex carbohydrates. Complex carbohydrates drip into the bloodstream over several hours, rather than suddenly flooding the bloodstream with sugar. The slow increase in sugars and the slow depletion extend energy over many hours and allow for a soft landing.

The best sources of complex carbohydrates are whole grains, such as brown rice, barley, millet, corn, and buckwheat; pulpy vegetables, especially the sweet root vegetables, such as squash, carrots, unions, parsnips, and rutabaga; and fruit.

If you eat a bowl or two of brown rice in the morning, you will have enduring energy until you are hungry at lunch. Be sure to add water and reheat your morning grain to make it moist and easier to digest. If you eat another whole grain at lunch, your energy levels will endure until dinnertime. Another grain at dinner will give you energy until you fall asleep at night.

I do not want you to eat only whole grains, however. I want you to enjoy the full spectrum of foods available to us, especially the vast variety of plant foods, including leafy greens, roots, round vegetables, fruits, and even some sea vegetables.

These long arcs of energy actually produce an entirely new energy wave within your system. They keep blood sugar levels up for many hours. You will have many hours of energy and feelings of strength, vitality, and confidence.

Here are some of the foods that will give you enduring energy for several hours after eating them. Include representatives from the following lists every day for long-lasting energy.

Whole Grains	**Sweet Vegetables**	**Beans**	**Fruit**
Amaranth	Acorn squash	Aduki	Virtually all fruit
Barley	Banana squash	Black beans	
Brown rice	Buttercup squash	Black-eyed peas	
Buckwheat	Butternut squash	Chickpeas	
Corn	Delicata squash	Kidney beans	
Millet	Hubbard squash	Lentils	
Oats	Hokkaido pumpkin	Navy beans	
Quinoa	Potato	Pinto beans	
	Pumpkin	Soybeans	
	Sweet dumpling squash		
	Sweet potato		
	Yam		
	Sweet root veggies such as carrots and parsnips		

One of the best ways to take you off the hypoglycemic roller-coaster ride is to eat a moist grain meal—by moist, I mean a grain cooked with adequate water—at least two times per day, along with a sweet vegetable. Make sure that the predominance of your diet is plant-based—that's where the complex carbs are located. Combine complex carbohydrates with small amounts of animal foods, and you will find that your energy levels rise quickly and stay up between meals.

I do not encourage the consumption of wheat, especially wheat bread, rolls, and pastries. Once the grain is cracked, it begins to decay. Many vitamins and minerals are lost as the bread sits on the shelf. Bread is rich in calories, and the carbohydrates in bread are rapidly absorbed. Both factors promote weight gain. In addition, many people suffer allergic reactions to wheat, wheat gluten, and wheat bread—both whole-wheat and white bread. Symptoms can include bloating, indigestion, constipation, headaches, fatigue, insomnia, depression, and excessive mucous production. If you suffer from any of these symptoms and eat wheat, especially in the form of bread, consider stopping bread, rolls, and other flour products. Finally, wheat is particularly high in phytic acid, which interferes with absorption of minerals, especially zinc.

Soaking grains, and throwing off the water, removes the phytic acid from grains and makes their mineral content, and that of the vegetables you eat with grains, more accessible to the small intestine.

Extending the Wave with Protein

Protein foods are slowly absorbed, meaning they keep glucose and insulin levels low. This will ensure that your body continues to burn fat. As I will explain in greater detail in the next chapter, protein also elevates the brain chemical dopamine, which increases feelings of energy, personal power, alertness, and aggression. On the Energy Balance Diet, you can eat a protein-rich food three times a day. After the first month, eat only one serving of animal food per day. This can include fish, lean beef, lamb, venison, pork, or eggs. The other two protein foods should be of plant-origin, such as beans, tempeh, or tofu.

Extending the Energy Wave with Breakfast

Lots of people believe that they are promoting weight loss by skipping meals, especially breakfast. Nothing could be further from the truth. What happens is that whenever you fast, even from one meal, you create more hunger and greater cravings, which usually results in a binge at some point in the day. When it comes to improving your health and achieving a healthier weight, you don't want to create excessive hunger. Once you are in the throes of hunger, you are more likely to eat indiscriminately, especially processed foods and foods rich in fat. That combination will wreak havoc on your energy wave, sending it into a hypoglycemic roller-coaster ride, with all of its emotional and physical reactions. Also, when you avoid meals, your body believes that it faces a famine and begins to slow metabolism in order to conserve calories. Rather than burning calories and causing weight loss, food deprivation actually causes your body to conserving calories and retain weight.

Instead of skipping breakfast, do the breakfast experiment I described earlier in this chapter in order to find the morning meal that's right for you. Whatever you choose to eat, combine it frequently with a moist grain; by moist, I mean add a little water and then heat so that the grain is soft and easier to digest. That will extend your energy wave and keep your energy levels up during the course of the morning.

People who skip lunch often find themselves bingeing at dinner. Again, skipping a meal intensifies the hunger and causes us to make bad choices later on in the day.

Snacking Between Meals

Contrary to what you might think, snacking between meals keeps insulin levels low—as long as you snack on a plant food, or a food that contains complex carbohydrates. Your first choice for a snack between meals should be a whole food, such as a vegetable, cooked grain, piece of fruit, or some nuts or seeds. Try quick-cooking vegetable soups—most can be prepared simply by adding boiling water—or a salad.

Choose foods that contain complex carbohydrates and are less calorically concentrated. Included in these are babaganoush (eggplant spread), hummus and rice cakes, salsa and rice cakes, or an unsweetened jam with rice cakes. Other suggestions can be found in Chapter 4.

Exercise Keeps Insulin Levels Low

Regular exercise burns blood sugar and fat, which is one reason why it keeps insulin levels low. Exercise also empties muscles of stored glucose. That's important because whenever you eat a processed food, or a food that contains sugar, the body can store the sugars that are entering the bloodstream. This gives the body another way of dealing with the sugars. The body does not have to burn all the sugars at once, but can store them, which means it can go on burning both sugar and fat. Regular exercise also promotes muscle growth. Muscle tissue is a big user of sugar and fat. The more muscle you have, the more sugar and fat you burn.

Taming the Sweet-Tooth Monster

A few years ago, a dentist came to me for help with a food problem. "All day long, I tell people that they shouldn't eat sugar," she told me. "It destroys their teeth and their health. But after I see two or three patients, I go into my back room and secretly eat candy bars. I can't stop myself. I want to stop eating sugar, but I can't do it. I'm addicted and I feel like an absolute hypocrite."

This was a very intelligent and sincere woman. It wasn't as if she lacked willpower or drive to achieve her goals. On the contrary, she was extremely successful and hardworking. If willpower alone was the key to success against sugar, she would have solved her problem a long time ago. But willpower is not enough when it comes to food dependencies, and especially a craving for sugar.

"You're not a hypocrite for wanting sweet flavor," I said. "You were designed that way. But here's what you can do about it."

I told her how to extend her energy wave by eating foods rich in complex carbohydrates and low-fat protein foods. I took her off all extreme processed foods. I specifically allowed her pasta and two processed products, rice cakes and rice syrup.

Rice syrup is a sweet syrup made from rice. It is processed, but contains an abundance of complex, as well as simple, sugars. Therefore, its effects on the body are much milder than standard candy bars, doughnuts, and other processed, sugary foods. Rice cakes are puffed brown rice. Yes, they are processed, but again they are rich in complex carbohydrates, which would help her further extend her energy wave.

"Buy a big supply of rice cakes and rice syrup," I told her. "Put it in your backroom and whenever you're craving something sweet, put rice syrup on a rice cake and eat that. Meanwhile, follow the program as closely as you can. The program provides foods that extend the energy wave, create balance, reduce cravings, and heal the organs (such as the pancreas) that are the source of many cravings. In two months, her sugar cravings had diminished remarkably. She found the rice cakes and rice syrup satisfying and adequate. Months later, I was told that she was urging her patients to eat rice cakes and rice syrup as a substitute for candy bars.

Deconstructing Cravings

One of the best ways to achieve balance and moderate the energy wave is by learning how to deconstruct your cravings and choose more balanced foods. Here's what I mean.

The driving force behind the extreme energy wave—and all that it creates, including weight gain and mood swings—are extreme foods, especially sugar and processed foods. But all extreme foods, including red meat and dairy products, cause cravings. Red meat and cheese trigger cravings for sugar and processed foods. Processed foods create extreme elevations and declines in glucose and insulin, which are the basis for an extreme energy wave.

It probably took you years to create your addiction and your extreme energy wave. It may take you a few months to heal yourself, though you will likely see changes within two weeks if you adopt my program. The cravings that you experience will be far milder, as well, after you adopt a more balanced way of eating. But there may still be some mild cravings. These can be managed easily and effectively. To help you do that, I have created a tool that I call "deconstructing cravings." By this I mean asking yourself a few questions and then listening and feeling for what your body is really desiring.

In order to effectively deconstruct cravings, it's important first of all to treat yourself and your body with a certain gentle understanding. Don't judge yourself for having cravings or a desire for any particular food, no matter how extreme that food may be. You are attempting to discover your own way of eating. Allow yourself a gradual transition from your current diet. Meanwhile, support your body with healthy foods and positive thoughts, even if you slip now and then and eat an extreme food. Treat your body well and it will be there for you, with all its marvelous talents, for a very long time to come.

Second, listen to your own inner desires with an entirely new level of concentration and openness to your body. Practice feeling where your tension is in your body and exactly what, if anything, you may be craving. The questions that follow will help you get the answers you are looking for.

Third, never forget the body's need for balance. Cravings originate when certain types of extreme foods dominate the diet. Extreme contracting foods create cravings for extreme expanding foods, and vice versa. The body's continual need for balance forces it to crave foods that will reestablish harmony. Therefore, whenever you experience yourself craving a certain food, reflect on what you've been eating that may have led to this craving. If you're craving expanding foods, such as sugary foods, perhaps you've been eating a lot of contracting foods, such as red meat, hard cheeses, salt, bread, or other baked flour products. If you crave meat, eggs, cheese, or pizza, check to see if you have been eating extreme expanding foods, such as sugar, pastries, soft drinks, and lots of fruit.

Recently, I found myself craving ice cream. I like ice cream, especially in the summer. But I don't like to eat ice cream more than a few times per year. Whenever I exceed my body's limit for ice cream, I suffer an array of health problems. So I only eat it when I feel my body really wants it and when I know I've been eating very well otherwise. That's when I can eat it without any lingering side effects.

My craving for ice cream emerged in the fall of 2001. The fall is not a good time to be eating ice cream, at least not for me. The season is turning cold, the weather is becoming more severe. Ice cream would only weaken my immune system and make me more vulnerable to a cold. So I began to wonder why that craving started to surface. What was I doing in my life that might trigger such a craving? I asked myself. At the time, I was teaching a lot of hours and my assistants were regularly bringing me hot tea. I drank the tea without thinking, but pretty soon I felt my body feeling hot and contracted. It wasn't long after that that I found myself craving ice cream. The stress of teaching is very contracting by itself, but I suspected that the hot tea might be adding to my contracted condition. Also the hot tea—which I was drinking in much greater quantities than I was used to—might be causing me to crave cooling foods.

"Please don't bring me hot tea anymore when I'm teaching," I told my assistants. "Instead, let me have some room temperature spring water." The next week I drank spring water. I also ate salads with olive oil. Raw salad is cooling and relaxing, and the olive oil provided a luscious, fatty satisfaction. Lo and behold, within a few days my craving for ice cream passed—without my ever needing to eat ice cream. (I look forward to eating it in the summer, though.)

I had achieved balance without turning to an extreme food, which ice cream certainly is, especially in the fall or winter. The lesson in all of this? Always look for the foods and behaviors in your life that may be the real source of your cravings. By changing those behaviors, you can reduce or eliminate the cravings, as well.

Finally, always remember that quantity changes quality. Your body can tolerate a certain limited quantity of an extreme food, without necessarily creating too extreme an imbalance. But when you exceed your limit, the results are usually suffering. Alcohol is a good example. A small amount will affect you very differently than a lot. It's the same with every other extreme food. Know and respect your limits.

How to Deconstruct Cravings

Whenever you crave something, reflect on what it is your body is asking for. Rather than grab for something that might approximate your craving, or just fill you up, focus on the characteristics of the food you are craving. Let's start with the flavor. Ask yourself: What flavor am I craving?

Are You Craving Something Sweet?

Believe it or not, there are many forms of sweet foods. There are chocolate, cookies, pastries, fruit, and fruit juice, just to name a few sweet foods. Also, there are cookies and pastries made with whole-grain flour and sweetened with fruit juice or barley malt. (For people with wheat sensitivities, be sure to avoid wheat products.)

As much as possible, try to satisfy your desire for sweet flavor with a milder, less extreme food. For example, try eating a rice cake with rice syrup or barley malt on it. You'll be surprised by how satisfying this treat is and how quickly it eliminates your need for extreme sugary foods. If something stronger is desired, try various cookies or pastries sweetened with fruit juice or barley malt, a sweet syrup made from barley. The idea is to substitute a less extreme food that closely approximates the sugary foods that you ordinarily turn to.

Quality makes a big difference in how the food affects you. If you decide to have an extreme food, choose the best quality you can buy. There's a good chance that you'll be satisfied with much less of that food. And then eat the food consciously, chewing it and thoroughly enjoying it. As long as you stay on the program, your cravings will diminish and you will soon find yourself being satisfied by more balanced foods.

Other possibilities for sweet desserts are provided in Chapter 10.

Are You Craving Salty Foods?

Cravings for salty foods often indicate a craving for minerals. Before you go out and have a bag of chips, eat a wide variety of vegetables, especially green and leafy veggies. This very often satisfies the craving for salty foods that in fact is a desire for more nutrition.

Are You Craving Bitter Flavor?

Bitter foods enhance digestion. Craving for bitter flavor is very often a craving for nutritious foods that cut through fat and stagnation in the middle organs and digestive tract, such as the liver and intestines. Most people satisfy their need for bitter flavor and digestive assistance by drinking coffee or dark beers.

If you find yourself craving bitter flavor, try eating bitter greens, such as dandelion, mustard greens, kale, and collard greens. These greens will unblock stagnant organs and promote healthy assimilation and elimination.

Are You Craving Pungent?

Chinese foods are often good sources of pungent flavor. Pungent flavor is another digestive aid. Grate fresh ginger on your vegetables or in your soup. In traditional Chinese medicine, ginger is an herb for the large intestine and lungs. It enhances function and promotes healing in both organs.

Are You Craving Spicy Foods?

Are you looking for an array of flavors, both subtle and strong, or are you looking for hot spices? So much of the American diet is lacking in flavor. It's been on the shelf a long time and is stale, bland, and tasteless. These foods lack vitality, energy, and aliveness. When fat and cholesterol are added to a diet of bland and stale foods, your body becomes overweight and stagnant. Blood becomes thick or viscous and circulation slows. As circulation weakens, organs and extremities become cool. At that point, the body may start craving spices.

When people start wanting spicy foods, they often turn to pizza or hot Mexican spices. These are extreme foods that do indeed warm the body, but create so much chaotic energy that they also stress the body. Instead of eating a pizza, with its dry, hard crust and heavy cheese, or refried beans and hot jalapeño peppers, try a bowl of linguine mixed with green vegetables and topped off with a nice marinara sauce that's got oregano, basil, onions, garlic, and celery. *Mangia*.

What Texture or Consistency Are You Craving?

First, consider that your condition might be overly tight and dry. Have you been eating a lot of bread, crackers, or other baked flour products? When eaten in excess, these foods create feelings of dryness and stagnation. They also make us feel stuck, hard, and irritable. When we reach that state of imbalance, we very often crave creamy, relaxing foods, such as ice cream, milk products, or oily foods.

Ask yourself if the creamy texture that you're looking for is more oil, such as olive oil? Or is it dairy food, such as milk or ice cream, that you're looking for? Try eating a salad with olive oil, or some pan-fried noodles or vegetables with olive or sesame oil. To pan fry, simply boil noodles and then put them into a frying pan with olive oil and sautéed vegetables. Delicious and very satisfying.

Are You Craving Something Moist or a Liquid?

First, ask yourself if you've been eating an excessive amount of salty foods or dry, baked flour products. Does your condition feel dry or tight? Are you thirsty?

I recommend that you try drinking pure spring water at least three times a day, preferably in the morning, afternoon, and evening. Put a bottle or a cup of pure spring water on your desk and sip it through the day. When the water enters your body, ask yourself how your body responds. If you suddenly awaken to how thirsty you are, your body is telling you that you've been neglecting your thirst. If you don't want the water, you will feel your body resist it, as if it were telling you that it has enough water inside and doesn't need any more.

Are You Craving Something Crispy and Dry?

If you are, maybe you've been drinking too many liquids. If that's the case, choose rice cakes or high-quality crackers without oil. Try to avoid chips. Even crackers are highly processed, which means they will elevate both glucose and insulin levels. Potato and corn chips are both rich in fats, especially saturated and trans fats, the two most injurious forms of fat. Both cause heart disease and promote cancer.

Are You Craving a Light or Heavy Food?

If you crave heavy foods, ask yourself if you've been eating a lot of salads and fruit. If so, are you also cold, especially in the hands and feet? Salads, fruit, and other raw foods make the body feel light. They also cool the body and can give rise to cravings for heavier, warming foods, such as fish, chicken, eggs, or beef.

If you crave heavy foods, choose a piece of fish over beef or hard cheeses. Fish is low in fat and high in omega-3 fatty acids, which boost immunity and prevent heart disease. Fish is rich in protein, as well. It's the healthier, less extreme, choice.

Are You Craving a Nutritious Food?

So often, I find myself wanting to eat something, and when I check with my body and discover what I am craving, the feeling I get back is that I want something of substance. What I mean is that I am looking for a nutritious snack. I find that this is especially the case when I am working hard and utilizing the nutrition my body is getting from my diet.

Are You Bored and Looking for Food to Entertain You?

We often use food as a distraction from our boredom. I go into this at length in the next chapter, "Power Foods for Brain and Body," but for now I am asking you simply to discover if that's the case. Learn to decipher true cravings from eating merely as a form of entertainment.

Are You Needing Exercise Instead of Food?

Another common craving from the body is for exercise. Tension builds up in the body throughout the day. Stress, hard work, and lots of thinking create that tension. The body sends us signals that it's in tension and is looking for a way to release its distress. Exercise is an ideal way of doing that. But all too often, we attempt to medicate ourselves with food as a way of dampening the tension and anesthetizing the body. Go out for a walk. I especially recommend that you adopt the exercise program described in the next chapter.

Physical health is the foundation of our lives. Once we free ourselves from extreme foods, the healing mechanisms of the body can focus their attention on overcoming our deeper physical and emotional issues. Then we begin to see healing miracles occur.

chapter 3

Power Foods for Brain and Body

If there is a single nutrient that people are struggling to find their own natural balance with, it is protein. People love protein. It makes you feel stronger, more alert, and more aggressive. In short, it increases your sense of power and confidence. All these things are good, which is one reason why high-protein diets are thriving today. Another reason people love high-protein diets, of course, is that they cause weight loss—and sometimes significant weight loss. Unfortunately, high-protein diets are rich in fat and cholesterol, and low in many essential plant-based nutrients, such as antioxidants, carotenoids, and phytochemicals. The high protein itself can also cause many serious health problems.

On the other side of the diet spectrum are the high-carbohydrate diets, which are low in fat, rich in plant-nutrients, but low in protein. If followed properly, these diets can lower your blood cholesterol, boost your immune system, protect you from many illnesses, and reduce your weight. Among the problems with high-carbohydrate diets is that their low-protein content makes them unsatisfying to many people and difficult to follow. What good is a diet if you cannot follow it for very long?

Also, many people create their own high-carbohydrate diets by including highly processed foods, which, among other things, increase weight and blood fats, known as triglycerides. Triglycerides are a risk factor in heart disease. About 20 percent of the triglycerides in your blood become blood cholesterol, which means they increase your cholesterol level and thus increase your risk of having a heart attack.

The big challenge we all face, it seems to me, is to find our own protein balance, within a plant-based diet. In other words, we have to create a program that combines the best of the two extremes, the high-protein and the high-carb diets. Somewhere in the middle lies the answer to good health, optimal weight, and a diet you can live with for life.

To find your own protein balance, you must know the impact of these two extreme diets on your health and weight, and how they attempt to create weight loss. If you know the dangers in the extremes, you can find the middle way for yourself. Let's have a look at these diets and what effects they have on people, starting with the high-protein programs.

Consequences of Filling Up on Animal Foods

Whenever a new group of students arrives at my school, I ask them about the kinds of diets they've attempted to follow. Inevitably, they've tried everything. Many have been on high-protein diets, by far the most popular form of weight-loss program available today. Others have been on moderate-protein diets, the so-called 30 percent protein, 30 percent fat, 40 percent carbohydrate diet. Still others have tried the many high-carbohydrate, low-fat diets, or the blood-type diet, or portion control diets, or various forms of vegetarianism.

Once I get a sampling of what people have tried, I ask about their experiences with these programs, beginning with the high-protein diets.

"How did you feel when you were eating the high-protein diet?" I ask. The following responses are common.

"I felt very tight," I recall one woman saying to me. "The diet created a lot of tension in me. It was as if my entire body felt squeezed."

"The more protein I ate, the more aggressive and anxious I became," another person said to me. "After a while, I didn't recognize myself."

"Constipated," said one, succinctly. "I had to take laxatives. I felt like my bowels shut down."

"I was always craving carbs—anything, a bagel, bread, pastries," a woman told me. "Every time I walked past the bakery aisle in my supermarket, or the donut shop near my apartment, I had to resist running in there and eating whole trays of doughnuts. I couldn't wait to lose the weight so I could stop eating that way."

"I couldn't stand eating all that meat, cheese, and fat," one man said. "It made me feel greasy."

At that point, I asked, "But did these high-protein diets cause weight loss?"

"Yes!" everyone said in unison. In fact, it's the rare person who has failed to lose weight on a high-protein diet.

"Were you able to keep the weight off?" I asked.

"No!" everyone told me.

"As soon as I went off the diet, I gained all my weight back. I even got heavier," one woman said. Other people in the class nodded their heads in agreement. She seemed to speak for everyone else who went on high-protein diets.

The fact is that high-protein diets do cause weight loss. There are two main problems with these diets, however. The first is that they cause many adverse side effects, some of which can be very serious. High protein consumption can damage kidneys and contribute to bone loss and osteoporosis. Also, high-protein diets tend to be high in fat and cholesterol, which contributes to heart disease. The second is that they cannot be followed for very long. Fortunately, most people cannot adhere to these diets long enough for them to do any real damage.

In fact, this is the case with virtually all diets. They can do for you exactly what they say they can do, but they can't be followed long enough to make a real difference in your life. High-carbohydrate, low-fat diets, for example, can cause weight loss and bring down your cholesterol level. They can also prevent many major illnesses, including heart disease. But many of those diets are so restrictive that you feel deprived most of the time. Portion-control diets often require that you limit the amount of food you eat, which means you have to be hungry much of the time. That can't last, as virtually everyone who follows such diets quickly finds out.

And then there's the fact that any diet in itself is foreign to your nature. They make you feel strangely out of sorts, off-balance, and different from the way you want to feel. The reason is simple: Food has a dramatic effect on brain chemistry. It very rapidly alters your energy levels and mood.

"It is becoming increasingly clear that brain chemistry and function can be influenced by a single meal," reported Massachusetts Institute of Technology (MIT) researcher John D. Fernstrom, Ph.D., in an article for the *Nutrition Action* health letter. "That is, in well-nourished individuals consuming normal amounts of food, short-term changes in food composition can rapidly affect brain function."

All of us use food to help us feel the way we want to feel. That means that diets are having profound and often unintended affects on both our physical and mental health. And, as I will describe, if the diet we are currently following creates feelings that are foreign to us, it becomes increasingly difficult to adhere to it.

Yet many people force themselves to stick to high-protein diets for as long as they can because they do cause weight loss. Here's how they do it.

High-Protein Diets—Too Much of a Good Thing

High-protein diets cause weight loss by tricking your body into thinking it's starving. They manage this trick by depriving your body of carbohydrates, its preferred fuel. Once your body recognizes that it is not getting carbohydrates, it thinks its starving. Its response to starvation is to shift into a condition known as ketosis.

Ketosis, essentially, is an emergency situation in which the body is forced to convert fat into ketones. Under normal circumstances, your body burns carbohydrates as its primary fuel. The body can also burn protein and fat, but that's not the case for the brain, which prefers only carbohydrates as a fuel source. When the body is deprived of carbs, it will burn protein for a few days. But evolution has taught us that burning protein is not a good idea. The reason: Protein is the primary constituent of muscle. Without muscle, you have no strength to hunt for food. Therefore, burning protein is counterproductive to survival. So the body doesn't like to do it, and it won't do it for very long.

Once the carbohydrate stores are burned, and you've burned protein for a few days, your body says, "It's time to start burning fat." The brain says, "No, I can't do that." At that point, the liver performs a little chem-lab magic and transforms fat into ketones, which the brain can burn. Now your body is burning fat and losing weight.

That's not the only way high-protein diets cause weight loss. During the first few days on the diet, you lose a lot of water, which accounts for much of the lost weight at the outset of the diet. High-protein diets also reduce hunger, which means you eat less. Finally, they keep blood sugar and insulin levels low, which ensures that you will keep burning fat.

Insulin, the Glycemic Index, and Weight Loss

As I described in Chapter 2, processed foods drive blood glucose and insulin levels upward. Once insulin rises, the brain tells the body to store the fat that's in your bloodstream, in your food, and in your tissues. It does this so that it can burn all the excess sugar in your bloodstream. Storing the fat is obviously bad for those who want to lose weight, which means you want to avoid high insulin. High insulin is also associated with adult-onset diabetes and with several types of cancer, including breast cancer.

Proponents of high-protein diets correctly state that the trigger for this increase in glucose and insulin is rapidly absorbed carbohydrates. Therefore the key to weight loss is simple, they say: Eat only those foods that are slowly absorbed.

The glycemic index is a test that shows the speed with which carbohydrates from specific foods are absorbed in the small intestine. Processed foods—such as bread, pastries, crackers, cookies, and sugary snacks—all have high glycemic index scores and therefore are obviously bad for weight loss. These foods, as listed in Chapter 2, are also rich in calories.

Unfortunately, many proponents of high-protein diets do not tell you that there's a big difference between processed carbohydrates—the bread, pastries, crackers, cookies, and sugary snacks—and whole, unprocessed carbohydrates from whole grains and vegetables. Whole grains, beans, and vegetables are slowly absorbed. They keep insulin levels low and keep you burning fat. These foods are also low in calories and fat. For the proponents of high-protein diets, a carbohydrate is a carbohydrate is a carbohydrate. Yet, even the glycemic index reveals that unprocessed carbs are more slowly absorbed than processed foods.

What the proponents of the high-protein diets and the glycemic index don't tell you is that fat will slow absorption. That means that a high-fat food looks great on the glycemic index, but very bad from a health standpoint. Fat is the most calorically dense substance in the food supply—a gram of it provides nine calories, as opposed to a gram of carbohydrate, which provides only four.

Foods that combine both processed carbs and fat are especially bad for weight and health, but because they contain fat, they are more slowly absorbed, which means they have a better glycemic index score than many low-calorie healthy foods.

For example, a Snicker's Bar—a food high in fat, sugar, artificial ingredients, and lots of calories—has a lower score on the glycemic index than a bowl of oatmeal. That means that, according to the glycemic index, you would be better off eating a Snicker's Bar than a bowl of oatmeal. It's worth noting that, ounce for ounce, the Snicker's Bar has about five times more calories than the oatmeal.

Still, because high-protein diets trigger ketosis, reduce appetite, and shed water pounds from the body, they cause weight loss. People think, "Oh well, the diet's working. I'm getting the results I want." They don't realize that they are also getting many unknown, and undesirable, results from their diet.

The Downside of High-Protein Diets

Once it is consumed, protein converts to acid in the body. The more protein in your system, the higher the acid levels. Your body works very hard to maintain a balanced acid-alkaline state within the blood. Once all that protein and acid come pouring into the bloodstream, however, the body must work fast to alkalize, or balance, the blood's pH. It does this by releasing its own natural alkalizing or buffering agent, namely the phosphorous in your bones. Unfortunately, when the phosphorus is released from bones, so, too, is the calcium. As the phosphorus and calcium are leached into the blood, the bones get weaker, or porous, creating the condition known as osteoporosis, or porous bones.

Meanwhile, in a highly acidic environment, other minerals are lost as well, including magnesium, which is essential for numerous metabolic functions, healthy bones, kidneys, and nerves.

Belinda S. O'Connell, M.S., R.D., reported in the health professional journal *Dietitian's Edge* (May–June 2001) that "We know that dietary protein intake influences urinary calcium losses with each gram of protein increasing urinary calcium losses by 1 [to] 1.5 mg. This means that a person who is consuming a high-protein diet requires more calcium in his or her diet to maintain calcium balance than someone who eats less protein. In situations where dietary calcium intake is sub-optimal, a high-protein diet may further worsen calcium imbalances and increase the risk of osteoporosis."

Harvard University's Mark Hegsted, Ph.D., reported in the *Journal of Nutrition* (1981) that doubling the protein intake increased calcium losses by as much as 50 percent. To anyone concerned about osteoporosis—and that should include virtually every woman living in the Western world—that information should be life-changing.

In places around the world where protein intake is low, or where it comes primarily from plant sources, rates of osteoporosis are exceedingly low. The Chinese are a good example. They consume 544 mg of calcium per day, on average. Americans, of course, are urged to eat between 800 mg (the minimum) to 1,200 mg per day, or higher, if you are at risk for osteoporosis. Yet, the Chinese have little osteoporosis. Why are their bones so healthy, even as they eat so little calcium? Because they also eat low levels of animal protein. The Chinese eat 64.1 grams of protein per day, on average. Sixty of those grams come from plant foods. Americans, by comparison, eat between 90 and 120 grams of protein, on average. Only 27 grams of their total protein intake come from plant foods.

As John McDougall, M.D., says in his book *The McDougall Program for a Healthy Heart* (Dutton, 1996), eating excess animal protein is the human body's "equivalent of acid rain." McDougall urges people to reduce their protein intake, because eating the amount of protein Westerns typically consume is like "pouring acid on your bones," McDougall says. (More about protein, calcium intake, and osteoporosis in the following section.)

In addition to creating more acid, high animal protein also converts to ammonia, a chemical that is highly toxic once it is in your bloodstream.

"Ammonia behaves like chemicals that cause cancer or promote its growth," says Dr. Willard Visek, Professor of Clinical Sciences at the University of Illinois Medical School. "It kills cells … and it increases the mass of the lining of the intestines. What is intriguing is that within the colon the incidence of cancer parallels the concentration of ammonia."

Nutritional Biochemist T. Colin Campbell, Ph.D., famed for his study of Chinese diets and health patterns, told *U.S. News & World Report* (May 20, 1991) that "Excessive animal protein is at the core of many chronic diseases. ... High protein intakes tend to decrease the cell-mediated immune response and tend to make one more susceptible to cancer."

Some scientists now believe that excessive consumption of protein may be as big a cause of cancer as dietary fat.

High-protein diets, of course, are composed primarily of animal foods, such as red meat, milk, sour cream, cheese, eggs, chicken, turkey, and fish. That means that you are eating a diet that's loaded with fat and cholesterol. That way of eating is associated with much higher rates of cardiovascular disease, cancer, adult-onset diabetes, high blood pressure, gout, and claudication, or inadequate circulation in the limbs.

Richard Fleming, M.D., of the Fleming Heart and Health Institute in Omaha, Nebraska, published a study in the October 2001 issue of the medical journal *Preventive Cardiology* showing that high-protein diets reduce blood flow to the heart muscle and increase the risk of heart attack.

But the fat and cholesterol content is only one of the problems of high-protein diets. Because they are so exceedingly low in plant foods, high-protein diets are also very low in antioxidants, carotenoids, phytochemicals, and fiber. When you adopt a high-protein diet, your body is suddenly deprived of the substances it uses to maintain the strength of its immune and cancer-fighting systems. The absence of fiber, coupled with the hard-to-digest sinew from meat, places enormous stress on your intestinal tract and causes constipation for many.

But the low fiber levels also mean that hormones are elevating. Fiber binds with hormones and takes them out of the system. High-fiber diets reduce estrogen levels dramatically. But without fiber, hormones remain in the system and can even become elevated.

Numerous health authorities have urged people to eat a diet that is very different from high-protein. The American Heart Association has urged people not to adopt high-protein diets, warning that they have many negative side effects, including adverse effects on the heart.

The U.S. Surgeon General stated in his review of the science on health and diet, titled *Healthy People 2000*, that people should increase their consumption of "complex carbohydrates and fiber-containing foods."

The Surgeon General urges people to eat at least five or more servings of vegetables (including legumes) and fruits, and six or more servings of whole grain products.

The National Cancer Institute (NCI) and the American Cancer Society (ACS) have issued similar statements.

Without adequate plant foods and carbohydrates in their diets, many people on high-protein diets experience a decline in health. They also suffer from acute cravings for carbohydrate-rich foods. These two side effects make it extremely difficult to remain on a high-protein diet. Eventually, the cravings for carbohydrates trigger a binge. People gorge themselves on bread, pasta, rolls, pastries, and doughnuts—anything to replenish the body's carbohydrate stores and satisfy its cravings. Of course, once a person eats carbohydrate-rich foods, he or she is immediately thrown out of ketosis. The body realizes that it is out of the starvation state and now burning carbohydrates again. But once out of ketosis, the person is gaining weight again.

When you are out of ketosis and gaining weight, you're probably also looking for another diet. Many people then go to the opposite extreme: a high-carbohydrate diet, or even a vegetarian regimen, such as that promoted by Dr. Dean Ornish.

The Other End of the Spectrum: Tasteless Diets

High-protein diets represent one extreme in eating, just as some of the high-carbohydrate diets do, especially the vegetarian regimens. The high-carb, low-fat diets might also be termed "low-protein" diets. These diets are 70 to 80 percent carbohydrates, which means they allow only small amounts of animal foods, if such foods are allowed at all.

The high-protein advocates have an effective weight-loss program that attracts adherents, at least for a short while. The high-carb, low-protein advocates have health as their primary argument. And indeed, the high-carb diets are a lot more compatible with the human body. The vegetables, grains, beans, and fruit provide the body with an abundance of vitamins, minerals, phytochemicals, and fiber, all of which the body desperately needs. Also, the low-protein diets place much less stress on the kidneys and bones and are associated with lower rates of osteoporosis. Still, many people find it impossible to follow a high-carbohydrate diet. There are some good reasons why these diets fail.

The Fear Factor

The first is that many people adopt high-carb diets out of fear of disease—and of food. For those who are ill and come to, say, the Pritikin Program, or Dean Ornish's program, the real motivation is often fear of another heart attack, or some other illness, or because they are overweight and cannot lose the weight. In other words, they come to the diet because of some negative incentive, and not because they love the food. Meanwhile, many never learn how to prepare the food to make it delicious and satisfying. They follow the diet plan for as long as they can, because they are afraid of what might happen to them if they go off the diet.

The fear factor can only travel so far. Eventually, many people on high-carb diets get frustrated from depriving themselves of the foods they really enjoy, and really want. At that point, they start cheating. They don't realize it at first, but cheating is a slippery slope. Pretty soon, they're more off the diet than on. And sooner or later, they throw up their hands and say, "I'd rather die five years sooner than eat that food." In other words, the fear factor lost its hold on them. While they were in fear, they never gave themselves a full chance to develop their cooking skills and to allow their taste buds to make a small adaptation to a healthier way of eating.

Adding Processed Foods to Get That Missing Taste

The second reason that these diets fail is because so many people invent their own version of the high-carb diet, and that version is based on processed foods, especially whole-grain breads, rolls, and muffins. Diet experts refer to whole-wheat bread and whole-wheat rolls as a "whole-grain food," but that's not true. These are processed foods, not whole foods. And unlike whole foods, processed grain products—even whole-wheat bread—are rapidly absorbed. They raise blood sugar and insulin levels. They also cause weight gain. In addition, many people have allergic reactions to wheat and don't know it. All they know is that they suffer from a digestive problem, or some other symptom, but do not attribute it to the bread or rolls or other wheat products they are eating. These are some of the reasons I encourage people to avoid bread and other wheat products, or, at the very least, eat them only sparingly.

That causes frustration, of course. People tell themselves that they're eating healthy foods. "I'm on a whole-grain and vegetable diet," they say.

But really, they're on a processed-grain and vegetable diet. Some of those processed grain foods—such as the muffins and pastries—also contain plenty of fat. The combination of processed foods and fat causes weight gain and health problems.

This has been one of the criticisms leveled against the type of program that permits processed foods as a substitute for whole-grain products. In addition to causing weight gain, processed foods also drive up triglycerides, or blood fats, which contribute to heart disease. That's another reason why we don't want our diets to be composed of too many processed foods.

The illusion that all grains are created equal is one of the big problems with the U.S. Department of Agriculture's Food Pyramid, which does not differentiate between a whole grain and a processed grain. The failure to make that distinction clear causes much consternation among people who are sincerely trying to do the right thing with their diets. They think they're eating foods that will cause weight loss and improved health, but they're wrong. (More about the USDA pyramid and why it's a poor guide for health choices in Chapter 9.)

Another important reason for the failure of many high-carb diets is that we in the West have not developed a culinary tradition around high-carb diets. The reason for this failure is due to how these diets arose within Western culture. High-carbohydrate diets have become a form of medicine in the West, and like medicine, they're not always the easiest things to get down your throat.

The Difference Between a Cuisine and a Hodgepodge of Foods

With the exception of macrobiotics, the purveyors of the modern high-carb diets created plant-based diets that were based on a single priority: The food had to be low in fat. That didn't mean that it was delicious or satisfying, however. In fact, most people find the high-carb diets tasteless, or worse.

If you look at traditional cultures and their plant-based diets—the Chinese, Japanese, Thais, and Mediterranean peoples are all good examples—one of the things that strike you is that their cuisines arose from a long tradition with food. They have been preparing the same foods for thousands of years. That means that they've learned how to use condiments, spices, and cooking styles to make the food delicious and healthy. All of these people use both plant foods and animal foods. They also include red meat, fowl, and fish. However, they use animal foods more as a condiment than as the center of their meal. The fact that these diets are low in fat and high in plant foods is more the consequence of nature—that's what nature provided these people—than it is of any intellectual design. The first priority of traditional people was to make the food that nature provided delicious and satisfying. In other words, these people created a cuisine that was meant to be enjoyed.

What we in the West have done is throw together low-fat foods from Mexico, Japan, China, and the great American food factory, and created a diet that is oftentimes a tasteless hodgepodge.

Also, we have been removed from natural foods for so long that we have lost a taste for vegetables, grains, beans, and fruit. That's why it's so important to learn how to prepare natural foods—because the only diet you will stick with is the one that you enjoy eating.

This is why macrobiotics and Mediterranean diets have enjoyed some success in the West—because they are based on healthful foods and a cooking tradition that give the food flavor and satisfaction.

The most healthful diets, such as those of Asia and the Mediterranean, are composed primarily of plant foods, with animal foods providing secondary sources of nutrition. But animal foods do play an important role in most traditional diets, including those of Asia and the Mediterranean. It's a safe bet that most people in Japan and Italy, for example, believe that their diets are satisfying and delicious. To the people who are eating these foods, the low rates of heart disease, cancer, obesity, and osteoporosis are secondary to the fact that they are delicious ways of eating.

You Have to Enjoy the Food to Succeed on the Diet

On my program, and in my recipes, I describe ways to prepare meals quickly and deliciously, using traditional condiments as a way of enhancing flavor. I stress that the food must be delicious and satisfying if you are to have any chance at sticking to the diet. A lot of the condiments and spices I use come from the Asian traditions, especially Japanese, and from the Mediterranean, especially Italy and the Middle East. The food must be easy to prepare in 30 minutes or less, as well. But I also urge you to learn how to prepare the foods so that you can be fully satisfied. Take a cooking class and learn how to make the grains, vegetables, beans, pasta, and fish and other animal foods delicious. Give your taste buds a little time to adjust, and use the spices and condiments that will make the food familiar and enjoyable.

But one of the crucial elements in discovering the diet that's right for you is to find out how much protein your body needs, especially from animal food sources. Everyone has different protein needs, for reasons I will describe later. So far, very few diets have offered both a high-plant food consumption and a low-to-moderate protein intake. That's the program I am offering.

In the United States, we have developed a diet that is centered around animal foods and protein, which is one of the reasons high-protein diets have been so successful. Americans have always had a love affair with protein and specifically with meat. Steak, eggs, and other animal foods have long been associated with power, masculinity, and success. In the minds of many Americans, vegetarians seem more relaxed, passive, and even a little flaky.

Those associations are derived, in part, from our intuitive knowledge of what animal foods and carbohydrate foods do to brain chemistry.

The Real Reason We Crave Protein

Let me start out by saying that there is no biological reason for us to crave the amount of protein—and animal foods—that we are eating today. Protein is used by the body for cell replacement and repair. You can get all the protein your body needs from a vegetarian diet. Moreover, all plant foods contain all the amino acids needed to produce complete proteins. (The belief that plant foods did not provide complete proteins was misinformation based on studies done on rats, not humans, back in the 1950s. Researchers have since cleared up those misconceptions.)

Researchers for the World Health Organization have established that men, nonlactating women, and children need to eat a diet that derives at least 5 percent of its calories from protein. Pregnant women should eat diets that contain 6 percent protein, and nursing women should get 7 percent.

You cannot help but get 7 percent of your calories from protein if you are eating a whole-foods diet. In fact, it's virtually impossible to avoid getting that much. Rice provides eight percent, corn 12, beans approximately 24 percent, a potato 10 percent, and most fruits at least 8 percent.

Actually, most Americans get six times the amount of protein their bodies actually need for basic metabolic function. I've already shown that excess protein is harmful to us. In fact, all that extra protein that we typically get is one of the primary reasons why we have such high rates of osteoporosis, or porous bones.

How Protein Makes Us Feel

The question is, Why do we want to eat all this protein? The real reason behind the protein craze is its effects on brain chemistry and how it makes us feel. Protein has a very different affect on our brain chemistry, attitude, and mood than carbohydrates do. Protein is the yang to carbohydrate's yin.

Researchers at the Massachusetts Institute of Technology (MIT), the University of California at Los Angeles (UCLA), and other research centers have discovered that brain chemistry, brain function, and mood can be altered dramatically by a single meal, usually within 10 to 20 minutes of consuming that meal.

Protein is made up of amino acids, which are the building blocks of protein. Individual amino acids have very different effects on brain chemistry. As I mentioned in Chapter 2, protein-rich foods, such as fish or meat, increase an amino acid called tyrosine, which in turn boosts brain levels of two chemical neurotransmitters, dopamine and norepinephrine. These chemicals create feelings of excitation, arousal, alertness, and aggression. Excess protein causes heightened states of arousal, stress, anxiety, and fear. Researchers have found that when dopamine levels are elevated and serotonin depleted, anxiety, fear, and depression are common.

"Dopamine and norepinephrine are the alertness chemicals," writes MIT's Judith Wurtman, Ph.D., in her book *Managing Your Mind and Mood Through Food* (HarperCollins, 1986). "Research done with laboratory animals and with human volunteers indicates that when the brain is producing dopamine and norepinephrine, distinct changes in mood and behavior take place. In people, these changes include a tendency to think more quickly, react more rapidly to stimuli, and feel more attentive, motivated, and mentally energetic. Problems, even big ones, often seem more manageable because of heightened 'brain power.'"

You don't need to eat a lot of animal protein to have this effect. "For most people, three to four ounces of a protein food delivers enough tyrosine to the brain to stimulate production of dopamine and norepinephrine, the alertness chemicals that keep you mentally up, thinking quickly, and accurately, with the brain power to tackle any challenges that come your way," writes Dr. Wurtman.

How Carbs Make Us Feel

The opposite effect takes place when you eat a carbohydrate-rich food, such as a whole grain or a pulpy vegetable. Carbohydrates increase an amino acid called tryptophan, which in turn elevates the neurotransmitter serotonin. When serotonin is elevated, you experience a greater sense of relaxation, calm, and inner peace. Your self-esteem rises, as does your sense of well-being. Serotonin also boosts your ability to concentrate. And when you're tired, it enables you to sleep more deeply and restfully.

At UCLA, Michael McGuire, an evolution psychologist, has found that leaders within groups tend to have higher levels of serotonin than the followers. He has studied athletes and found that team leaders have higher-than-average levels of serotonin in their brains compared to their team members. Similarly, McGuire has found that fraternity officers also have higher brain levels of serotonin than their fraternity brothers.

These findings have been consistent throughout the animal kingdom, McGuire has found. Monkeys who are the leaders within their societies have much higher levels of serotonin than those at the lower end of the social ladder.

Serotonin is responsible for maintaining elevated mood and optimism. When serotonin levels fall, depression is usually the result. As you probably know, most of the anti-depressant drugs used today, such as Prozac and Zoloft, are known as serotonin uptake inhibitors. These drugs work by increasing brain levels of serotonin.

Serotonin is elevated rather quickly by any carbohydrate-rich food. "If you eat a carbohydrate alone, without protein, more tryptophan will be made available to your brain, which will use it to make more serotonin," writes Dr. Wurtman. "As a result, you will feel less stressed, less anxious, more focused, and relaxed."

Because carbohydrate foods have these effects, Dr. Wurtman has labeled them calming foods. Indeed, "feelings of stress and tension are eased and the ability to concentrate is enhanced" after eating a carbohydrate-rich meal, reports Dr. Wurtman.

"Those who eat calming foods report feeling more relaxed, more focused, less stressed, less distracted after their meal," Dr. Wurtman has reported. Research has found that people tend to score better on tests that require concentration and focus after eating a carbohydrate-rich meal.

Remarkably, the changes brought about by eating a carbohydrate-rich food or a protein food occur rapidly—usually within 10 to 20 minutes of eating. And it doesn't take much food to create these changes.

Managing Our Mood with Food

We are all trying to manage our brain chemistries, and our moods, with our daily diets. We intuitively know how food affects us. Many people turn to steak or eggs when they feel they need more power. They use sugar or alcohol—a form of refined sugar—to relax, cool out, and let go of the tension of the day. In this way, we are all using proteins and carbohydrates to create the internal effects that we are looking for.

One of the things that many people want—especially those who are ambitious and live in cities—is a sense of personal power, control, and a rapidly functioning mind. Hence, they crave protein in the form of animal foods. This is one of the reasons why protein is so desired today: People want to feel more powerful, more in control and in charge of their lives. But when too much protein is consumed, dopamine and norepinephrine dominate brain chemistry, causing anxiety, fear, nervous tension, and even paranoia.

These extreme feelings can be balanced and eliminated by reducing protein intake and increasing carbohydrates. With consumption of grains and vegetables, your sense of calm, well-being, optimism, and self-confidence are restored.

In today's culture, most people are eating excesses of protein and trying to balance it with sugar and processed foods. These foods give an initial serotonin rush that is soon followed by a drop-off in serotonin levels. Sugar boosts serotonin almost instantly, and, thus, creates an almost instant state of well-being and relaxation. (This is one of the reasons why people are so addicted to sugar today.) However, sugar and other processed carbohydrates are rapidly burned off, causing a precipitous fall in blood sugar and serotonin levels. Once blood sugar starts falling, so, too, do serotonin levels. This is what causes the depression, feelings of fatigue, and loss of confidence so often associated with hypoglycemia.

As serotonin levels fall, dopamine levels remain unchanged. In fact, they are now excessively high relative to serotonin—which creates feelings of anxiety, stress, physical tension, and emotional distress. These are also common feelings associated with hypoglycemia.

The energy wave can now be seen as having two components: blood sugar and serotonin. As long as both of these remain consistently up, you experience higher energy and a good mood. But when they fall, as they do shortly after we eat sugar and processed carbohydrates, we experience a drop in energy and mood, as well as all the other side effects associated with hypoglycemia.

The best way to keep the energy wave up is to eat whole, unprocessed grains and vegetables, which create long arcs of blood sugar and serotonin levels. This is why a diet that is dominated by plant foods—otherwise known as serotonin boosters—and is coupled by the amount of protein that's right for you, will give you long-lasting energy, enduring feelings of well-being, and a sense of personal power and strength.

Experiment with Protein and Listen to Your Body

You cannot find that balance unless you listen to your body. If you follow the program I describe in Chapter 4, you'll be able to manipulate your protein intake to discover what feels best for you. Experiment with the amount of protein you need. Also, now that you realize what protein can do for brain chemistry, try to eat protein on days when you need a little extra assertiveness and alertness. Eat a protein-rich meal when you attend a power lunch, or an important meeting where you will be asked to speak. On the other hand, reduce your protein consumption when you need to concentrate or study for several hours or more.

I urge you to keep track of the foods you eat by keeping a diet diary. In that diary, note how you feel after your meals—are you feeling more alert, more aggressive, more mentally active? If so, your dopamine levels are elevated. In all probability, you just ate a protein-rich meal. If you are feeling more calm, relaxed, better able to concentrate, your serotonin levels are up. In that case, carbohydrates are dominating your diet.

Whenever you want to feel more powerful, more alert, and mentally active, increase protein. This is especially the case when you face some particular challenge, such as an important business meeting or some task that needs to be tackled.

When you want to study, concentrate, or relax, eat a carbohydrate-rich meal, such as brown rice, barley, amaranth, or quinoa, and any of the sweet vegetables. These foods will elevate your serotonin levels and keep them elevated for hours.

The Energy Balance Diet will balance both your dopamine and serotonin levels so that you can experience both the power of protein-rich foods and the calm optimism and self-confidence of the serotonin boosters. The result will be a deeper experience of balance and a growing sense of your own inner self.

Blood-Type Diet to Help Find Your Protein Needs

As I have been stressing throughout this book, one of the keys to finding your own healing diet is to discover how much protein is right for your body. One of the best guides I have found to discovering your protein needs is to use the Blood-Type diet as a general guide for experimentation.

In his book *Eat Right for Your Type* (Putnam, 1996), Dr. Peter D'Adamo states that as humans evolved from the primitive hunter-gatherers to agrarian, farm-based communities, our blood type changed. Each major stage in our evolution was associated with a different way of eating, and with a different blood type. At the hunter-gatherer stage, which is associated with the emergence of modern humans, people ate a combination of wild plants—such as roots, tubers, berries, and nuts—and meat. Meat, D'Adamo says, was the primary food of these people. Over many millennia, hunter-gatherers adapted genetically to this meat-and-wild-plant-centered diet. The blood type that emerged among these people was O. That blood type represented not only the blood, however, but also the nature of the immune and digestive systems. The entire biological systems of people with O blood function best on a meat-centered diet, D'Adamo maintains.

Because hunter-gatherers existed long before the emergence of agriculture, their systems never adapted to the primary foods from the farm, especially certain grains, such as wheat, and milk products from cows. Neither of these foods were consumed by hunter-gatherer societies. Today, people with O blood still retain the genetic makeup of their hunter-gatherer ancestors, D'Adamo asserts. Consequently, their immune and digestive systems work best when their diets are made up largely of flesh foods, such as meat, poultry, and fish. On the other hand, they find wheat and dairy products extremely difficult to metabolize. Consequently, both of these foods create biological imbalances, overweight, and sickness, D'Adamo says.

About 15,000 years ago, agriculture emerged and people began to produce their own food; they also domesticated animals. The diet changed dramatically, going from a meat-centered regimen to a plant-based diet composed largely of grains, vegetables, and beans. Also included in this diet was the milk from domesticated cows. D'Adamo maintains that the genes of these agrarian people adapted to this new way of eating, which produced the blood type A. That blood type, along with the alterations in the immune and digestive systems, functioned best on a plant-based diet. This is why people with A blood are best suited for vegetarianism.

Nomadic tribespeople emerged later and consumed a diet that was more a combination of farm-produced plants and animal foods. This blood type had a more balanced and flexible approach to eating, which permitted people to eat both plant and animal foods. B types are better able to consume dairy products, as well, D'Adamo asserts.

My experience with this diet, as well as the experiences of the students at my school, has shown me that there is a great deal of truth to it. People with type O blood tend to be more physically oriented, as hunter-gatherers once were, and have greater demands for protein. Also, O blood people often have difficulty digesting and metabolizing wheat. People with A blood are often attracted to vegetarianism. B types do a little better than others with dairy foods.

But just as with all the other dietary programs, this one also takes its basic premises to the extreme. People with O blood do not require the amounts of animal foods that D'Adamo recommends. Beyond a certain point, protein is injurious to everyone, regardless of blood type. It is among the most important reasons that the West suffers from such high rates of osteoporosis, digestive disorders, heart disease, and various forms of cancer, including breast, prostate, and colon cancers.

In addition, we have all evolved to become dependent upon whole grains, even the O blood types, though they typically do better on brown rice, millet, quinoa, and amaranth, the grains that have less gluten than wheat and barley. Grain is the staple for every civilization, all of which have included O type people.

Still, the D'Adamo perspective deserves to be considered when it comes to determining how much protein you should be eating. People with O blood should experiment with more protein, while maintaining a plant-based diet. People with A blood should consider eating less protein, and perhaps restrict that protein to fish only. B types can eat more protein than A and less than O. The diets of all three types should be based on plant foods, however.

In Chapter 4, I provide specific instructions on the consumption of animal foods. I also show you how you can use the D'Adamo system as a guide to eating more or less protein-rich foods.

The one protein food category that needs special attention today, however, is dairy products. These foods are consumed in much greater quantities than is healthy for anyone, no matter what your blood type may be. Dairy foods have become the silent source of illness in Western cultures today. Very few people realize just how much damage these floods are doing to us, especially to the young. Let's have a closer look at how dairy products are causing so many health and weight problems among us.

We Are Not Baby Cows

Occasionally I will ask my students what the problem is with dairy products. Many of them are well informed and consequently the list is fairly long. Invariably someone says that dairy foods cause mucous.

"How do you know?" I usually ask.

"When I stopped eating dairy foods, I stopped having a runny nose and a lot of mucous discharge," the person tells me.

"It's not scientifically proven that dairy products increase mucous production," I reply.

"Yes, but it worked for me," the person says. "When I stopped eating dairy, the mucous stopped."

"Okay," I say, "then it is scientifically proven—for you. You did an experiment and found out that dairy foods clog your body with mucous. So you know firsthand what dairy products do for you. You don't need someone else to tell you that you can eat them or you can't eat them."

Women who have experimented with dairy foods inevitably tell me that these foods cause difficult, painful periods and premenstrual syndrome, or PMS. When I ask them how they know, they tell me that when they stopped eating dairy foods, their symptoms went away.

"So you're telling me that you have derived this information from your own experience?" I ask.

"That's right," they say. They have found that their bodies react in this way to cow's milk and products derived from it.

I stress the importance of personal experience because experts can dredge up scientific studies to tell you that a particular food is either good or bad, depending on your personal persuasion. Contrary to what many people believe, however, the scientific evidence is growing rapidly to show that dairy products are bad for all of us, especially children.

The truth is, cow's milk is the perfect food for turning a 65-pound calf into a 500-pound animal in less than a year. Unfortunately, it's doing something very similar to humans. It also causes many serious health problems. Here are just a few.

Dairy products are rich in fat and cholesterol, including saturated fat, which causes heart disease. Even when the fat is removed from milk products, however, it still causes a whole range of serious disorders, including cancer and diabetes.

Numerous studies have shown that the sugar in dairy products, lactose, injures the ovaries in women and may lead to ovarian cancer. Daniel Cramer, M.D., and his colleagues at Harvard University, found that, once consumed, lactose is transformed into another sugar, galactose. The body responds to galactose by producing enzymes to metabolize and eliminate the sugar. But when excess galactose is ingested, it exceeds the body's ability to produce enough enzymes to get rid of it. The galactose builds up in the tissues and adversely affects a woman's ovaries, Cramer reported in the medical journal *The Lancet* (1989; 2:66–71). Women who eat dairy products on a regular basis have three times the occurrence of ovarian cancer that women who avoid dairy foods do, according to Cramer's study.

Before it triggers a cancer, however, the galactose very likely causes a significant amount of distress in a woman's reproductive system, including, perhaps, PMS and painful periods.

A growing body of evidence is showing that dairy products contribute to both breast and prostate cancers. Consumption of cow's milk causes a rapid increase in insulin-like growth factor (IGF-1), high levels of which have been linked to both breast and prostate cancers. Men with the highest levels of IGF-1 have four times the incidence of prostate cancer, compared to those who do not drink cow's milk and have low levels of IGF-1.

The proteins in milk products cause an autoimmune reaction that, in sensitive children, causes incurable juvenile diabetes. A study published in *The New England Journal of Medicine* (July 1992) showed that the proteins in milk products (specifically bovine albumin peptide) attach themselves to the insulin-producing cells of the pancreas. The immune system recognizes these proteins as a threat to health and attacks them. Unfortunately, it also destroys the beta cells of the pancreas, which produce insulin. Populations studies from around the world have shown a strong correlation between the consumption of dairy products and insulin-dependent diabetes.

Children appear to be particularly at risk from dairy consumption. More and more research is showing dairy foods are associated with a laundry list of problems in children, including digestive disorders, iron-deficiency anemia, asthma, and allergies.

Most of the world's population, in fact, is lactose intolerant, meaning that most humans lose the ability to digest lactose. This includes most Asians, Africans, Mexicans, and Native Americans. In fact, about 15 percent of Caucasians are lactose intolerant, too. People who are lactose intolerant and who eat dairy products suffer from a wide array of symptoms, including gastrointestinal disorders, diarrhea, indigestion, heartburn, and flatulence.

Milk products are laced with the antibiotics and recombinant bovine growth hormones given to dairy cows. Milk routinely contains pesticides that the cows consume in their food. Those pesticides are concentrated in the animal's fat cells and transferred to its milk.

Many people believe that you need dairy products to protect yourself from osteoporosis, but the research has shown exactly the opposite.

Populations studies from around the world have shown the countries where dairy foods are consumed have the highest levels of osteoporosis.

In an article for *Science* (April 22, 1994), the chairman of Harvard University's Department of Nutrition, Dr. Walter Willett, pointed out that the high levels of protein in milk promote the loss of calcium through the kidneys. In an article titled "Diet and Health: What Should We Eat?" Dr. Willett wrote that "adult populations with low fracture rates generally consume few dairy products. … [Dairy products] contain a substantial amount of protein, which can enhance renal [kidney] calcium loss."

Other research has supported this finding. The Harvard Nurses' Health study, an enormous research project that has followed the health of 80,000 nurses for more than 12 years, has found that regular milk consumption gave no protection against osteoporosis. On the contrary, the increased protein and calcium from milk is associated with a higher rate of fractures, according to the Harvard researchers. Other studies have corroborated this finding.

"The myth that osteoporosis is caused by calcium deficiency was created to sell dairy products and calcium supplements," said John A. McDougall, author of many books, including the *McDougall Program for Healthy Women* (Penguin Dutton, 1998). "There's no truth to it. American women are among the biggest consumers of calcium in the world and they have one of the highest levels of osteoporosis in the world. Eating even more dairy products and calcium is not going to change that fact."

Whenever I start talking about dairy foods, people inevitably panic and say, "Where am I going to get my calcium?" There are many excellent and healthful sources of calcium.

Calcium is abundant in green vegetables, as well as many other sources. A cup of milk contains about 300 mg of calcium. Here are some alternatives to milk as a source of calcium.

- A cup cooked collard greens contain 360 milligrams of calcium.
- A cup cooked kale contains 210 mg.
- One cup cooked bok choy contains 250 mg.
- One cup fresh broccoli contains 140 mg.

Note: All dark green and leafy vegetables contain significant amounts of calcium; the darker the green, the more calcium and other nutrients in the vegetable.

- Four ounces tofu (about the size of a deck of cards): 150 mg.
- One cup cooked beans: 100 mg.
- Nuts and seeds, including almonds, walnuts, and sesame seeds, are all good sources of calcium. (Almonds have about 130 mg in a third of a cup; sesame seeds about 80 mg per half cup. In addition to their use as a snack, seeds can be eaten as a condiment on grain, such as brown rice.)
- One tin sardines: 480 mg.
- Three-and-a-half ounces salmon: 290 mg.
- Three-and-a-half ounces mackerel: 300 mg.
- Three-and-a-half ounces herring: 250 mg.
- Mineral water, which often contains significant amounts of calcium and is easily assimilated by the body.
- Sea vegetables or seaweed. See recipes in Chapter 10 for ways to prepare sea vegetables.
- If you feel strongly that you need additional calcium, you can take a calcium supplement. It will have a less harmful effect on your body than drinking milk or eating milk products. If you take a calcium supplement, be sure it contains magnesium. Calcium cannot be absorbed without magnesium. Also, calcium without magnesium places enormous stress on the kidneys in order to excrete the excess calcium from the body. This can result in damage to the kidneys and kidney stones. Anyone who takes calcium without magnesium is probably doing more harm than good.

In the next chapter, I give specific recommendations for how you can experiment with dairy foods to discover if they are causing you to suffer health problems. I also show you how you can reintroduce them in smaller amounts if you desire dairy products.

chapter 4

The Energy Balance Program

Finding and maintaining a healing diet that's right for you is one of life's greatest achievements. Think about it: How many of your accomplishments can radically improve your health, protect you from major illness, give you abundant energy, help create and sustain a positive outlook, and make you look and feel years younger than your chronological age? When you consider all the suffering, medication, and surgery that you can avoid by eating health-promoting foods every day, you start to realize that a healing diet may be one of the four or five greatest gifts you could ever receive.

The challenge, of course, is that you must find it for yourself. You earn the gift of good health only after you have searched for it. Today, there are millions of people searching for that gift. They are buying diet books and listening to lectures hoping that this latest program is the answer. What people don't realize is that the best a program can offer is a foundation and a compass to finding your own unique way of eating. The foundation is a basic, health-promoting diet, to which you can add foods according to your needs. The compass is the understanding you need to make adjustments in your way of eating. As I have been stressing throughout this book, the compass must be an intimate connection you make with your inner wisdom and balance, which is the best guide to your nutritional, psychological, and spiritual needs.

The program that I provide in this chapter can help you establish both of these things in your life. The way of eating that I recommend can give you a direct experience of your own inner balance and a clear sense of how individual foods affect you. By knowing how food affects you, you can create a way of eating that is designed strictly for your unique health condition and constitution. Such knowledge can lead you to good health and to freedom.

You don't have to have that knowledge immediately. On the contrary, all you have to do for now is adopt the foods on the Energy Balance Diet, which I describe in the following sections. As you follow this diet, you can experience a remarkable improvement in your health. Among the benefits that this program can give you are the following:

- Tremendous increase in vitality and overall energy supply.
- Weight loss, if you are overweight, and the ability to maintain your healthy weight without starving yourself or living an uncomfortable and restrictive lifestyle.
- A more youthful and healthy appearance, including more beautiful skin.
- Greater clarity of mind and improved memory.
- Much-improved digestion and elimination.
- Deeper and more restful sleep.
- Greater optimism and self-confidence.
- Relief from many chronic symptoms and conditions.
- An experience of your own inner balance, as well as a greater awareness of how food affects you.
- A way of eating that can help you restore balance whenever you feel weak, pessimistic, or ill.

Seeking Balance

People often ask me, What is balance? What should I be looking for or experiencing? Balance is a state in which you have a powerful experience of your own physical, emotional, and psychological integrity. You will have greater vitality, enduring energy, and a certain lightness of being. You will feel very much in touch with yourself and very grounded in your own body and experience of life. In a place that is nonverbal and entirely experiential, you will have a more direct knowledge of your life and yourself. You will need food, but you will not be overpowered by intense cravings. As you experience your own state of inner equilibrium, your cravings for all sorts of things—including the intense neediness that we sometimes feel for other people—will become less intense. Rather, you'll enjoy your own company more—and, ironically, the company of other people as well. You'll experience greater peace, relaxation, and joy in your own state of being. You'll feel contained within yourself and your own center, yet you will be perfectly capable reaching out to people and connecting with those you truly care about. That's balance.

The miracle is that certain foods, and a certain type of diet, can lead you to this state of balance. That diet is described later in the chapter.

As you follow the diet I have described in this chapter, you will grow increasingly into this state of balance. The process is quickened by your increasing awareness and sensitivity to what food is doing to you—physically, emotionally, and psychologically. For example, you'll know for sure how sugar affects your body, mood, and thinking. You'll experience how processed foods affect your digestion, energy levels, and mood. On the other hand, you'll know the strength, vitality, and equilibrium that vegetables, beans, and grains give you. No one will have to tell you how these and other foods affect you. You'll know. In this way, you will be able to create your own limits, and your own diet. You'll also know how to restore balance and health when you have gone off your diet and feel weak or ill. The latter accomplishment takes time and experience with the basic Energy Balance Diet.

At my school, the Institute for Integrative Nutrition, we spend six months with students evolving each person's diet so that he or she finds his or her own right approach. Some people find the diet that's right for them within the first month. Others take longer. But everyone learns to adapt their way of eating to fit their circumstances and ever-changing needs. I mention this because I want you to know that finding the diet that's right for you takes a little time, experimentation, and adaptability.

Follow the program I have designed. This will lead you to health and a greater awareness of how food affects you. That combination—health and awareness—gives you the basis for appropriate exploration of the world of food.

Let's Get Started

I designed this program for people who live busy lives and who have limited amounts of time to prepare food each day. You can come home and prepare virtually all the foods recommended in this program in 30 minutes or less. You can achieve all of your health goals in this way of eating—and do it within your real-life schedule.

The Energy Balance Diet is designed to gradually incorporate balanced and healing foods into your diet. I have found that a sudden, shocking shift in diet makes it more difficult for people—especially essentially healthy people—to maintain a new diet. The reason is that a sudden shift in diet forces you to discipline yourself and repress your food cravings. The longer you repress your cravings, the more powerful they become, until you finally relent and the diet fails. A gradual introduction of balanced foods, on the other hand, allows several important changes to take place without much effort.

First, it gives the healing foods time to change your biochemistry and to create greater balance within your system. This alone reduces your cravings for extreme foods. Among the changes that this gradual shift allows is for your palate to adapt to the new flavors.

It also gives you time to learn how to prepare the foods to make them delicious and satisfying. You'll also need time to get used to shopping for and preparing the foods. Chapter 10 provides more than 100 delicious and health-promoting recipes to help you make the transition to a healthy way of eating. In a very short time, you will not feel any sense of deprivation. Your new diet will be as satisfying as the one you are eating now; the big difference, of course, will be that you'll feel healthier, happier, more energetic, and mentally clearer.

Finally, the gradual introduction of balanced food into your diet allows these healing foods to crowd out the extreme foods that you are currently eating. As your body and palate adapt to the new foods, and as you learn to prepare them to your satisfaction, you find that the extreme foods are gradually disappearing from your daily way of eating. Very soon, you realize that your balanced diet is your normal way of eating, something you don't even think about anymore.

Only you can determine just how fast you can fully incorporate the recommendations I have outlined in the following sections. Take your time. Make the changes at your own pace. But never let go of your goal to establish a healing way of eating, and to experience all the benefits that such a diet can give you.

The Short Course in the Energy Balance Diet

The Energy Balance Diet is composed of four main food groups and a few supplemental elements. The four main parts of the diet are as follows:

- **Green and leafy vegetables.** Eaten two to three times per day. These include broccoli, cabbage, collard greens, kale, mustard greens, and watercress.
- **Sweet vegetables.** Corn, carrots, beets, parsnips, squash, sweet potatoes, and yams. Sweet vegetables are eaten daily.
- **Protein foods.** The protein foods include animal foods, such as beef, eggs, chicken, and fish, as well as beans and bean products. Eat protein foods two to three times per day.
- **Whole grains.** These foods are eaten cooked at least once a day and preferably twice. Whole grains include amaranth, quinoa, brown rice, barley, oats, and pasta.

The supplemental categories include the following:

- Health-promoting snacks, soups, and desserts.
- Condiments, oils, and spices.
- Foods to reduce and foods to avoid.

I do not expect you to adopt the diet all at once. Rather, I recommend a step-by-step approach in which time is taken for each step to be integrated into your life. In most cases, a week is enough time to discover how the new food that you are introducing into your diet is affecting your body and overall health. Obviously, for those who wish to adopt the diet immediately, and *in toto*, you are free to do that.

The program begins with two simple but essential steps.

Steps 1 and 2

Drink more pure spring water and increase your awareness of what you eat. Keep a food and symptom diary in order to help you promote greater awareness.

Drink Pure Water

For one week, make only two small but important changes in your way of life. First, drink more pure spring water or filtered water. I recommend that you drink water in the morning, afternoon, and early evening. The amount of water you drink depends on how your body responds to the water during your first few sips. Sometimes, I drink a little water and find myself satisfied. I don't want any more, so I stop drinking. But very often, I drink a few sips and my entire body awakens to its thirst and need for pure, clean water. When that happens, I drink until I know my need for water has been satisfied.

You can live for a month without food, but you cannot survive longer than two or three days without water. Water is essential for your life. Most of us have not experienced a severe drought, so we don't have the first-hand knowledge of just how important water is to our survival. But our bodies need water to conduct every biological function. Many people experience poor digestion, sluggish thinking, and fatigue because of a lack of clean, pure water. Pure water promotes healthy bowel function, as well as elimination of toxins and waste products from the skin and urine. Water replenishes cells and restores relaxation to muscle tissues.

Whenever we are under stress, or hurrying through our day, one of the first things we become unconscious of is our thirst and our need for water. In order to give the body all the water it needs, we very often have to make a habit of drinking water. Therefore, I recommend that you consciously drink water three times a day—morning, afternoon, and evening. Keep a bottle or cup of pure water on your desk and sip it throughout the day. Put a bottle of pure spring water on your nightstand and drink when you wake up in the middle of the night feeling parched and thirsty.

Become Aware of How Food Affects Your Body

The second action to take during the first week is simply to become more aware of how food affects your body, mind, and emotions. Eat consciously.

Chew your food thoroughly and feel how it affects you when it's in your mouth and in your stomach. After you have finished your meal, consciously tune in to your body and become aware of the food's impact on your digestion, nervous system, and heart.

How is it affecting your overall vitality, heath, and mood? Do you feel stronger after the meal or weaker? What cravings do you experience, if any, an hour or two after your meal? Do you feel light or heavy from your foods? Are there any specific symptoms that arise from the foods you ate? What is your mood after you've eaten? How about an hour or two after the meal?

Keep a Food and Health Diary

To help you become more aware of what you are eating and how foods affect you, I encourage you to keep a daily diary of the foods you are eating, and any symptoms that may arise during the course of the day. Note any cravings that occurred during the day. Also, keep track of how severe the craving was. What was your reaction to the craving? Did you eat the food you wanted, or did you abstain? How did you feel after you ate the food that you really wanted? Or how did you feel by abstaining from it? In either case, did the feelings change an hour or two later?

It's very important to note any symptoms that arose during the day, and the times at which the symptoms emerged. Don't worry if you can't link the foods to the symptoms. All I want you to do is become aware of your food choices, cravings, and any physical or emotional discomfort, irritation, or pain that you may be experiencing. I also want you to create a baseline record to see how your dietary changes may affect these cravings and symptoms. A food diary is one of the tools that will promote rapid and healthful change in your way of eating, as well as your overall health.

Make the Change That You Know You Can Make

Very often when I am counseling someone, the person already knows that he or she should make a change in his or her diet. Many people are addicted to coffee and know the effect coffee is having on their bodies. When they come in to see me, they say, "I'm a coffee addict. I should give it up."

"Okay," I say. "Let's start with that."

Very often, we already know the first steps we should be taking in our return to health and balance. I urge you to take those steps now. After you have done that, become aware of how that change affects your health, energy levels, mood, and clarity of mind.

Step 3

Eat green and leafy vegetables two or three times per day.

One of the most important food groups that is missing from the diets of most Americans is green and leafy vegetables. Green and leafy vegetables provide an almost immediate experience of freshness, lightness, and strength—the latter benefit comes from their rich nutrient content, no doubt. They are rich in fiber, which promotes healthy elimination and the balancing of hormones. (See Chapter 8 for more on hormonal health.) The abundance of vitamins, minerals, carotenoids, and phytochemicals make leafy and green vegetables miracle foods, in my view.

Don't worry about serving size. It doesn't matter how much you eat of any vegetable—they're so low in calories that they will not add weight, though they will fill you up, thanks to their fiber and water content. The important thing is that you eat these vegetables at least two or three times a day.

Recommendations

- Eat at least two servings of cooked green or leafy vegetables per day. If possible, eat three servings.
- You can eat the same green vegetable twice and count it as two servings.
- Do this for at least one week before adopting the next step in the program.
- Become aware of how these vegetables affect your body and overall health.
- In your food and health diary, continue to report any changes in your cravings and health while you adopt the green vegetables.

In the following list, I have highlighted the vegetables that I particularly urge you to eat with three stars, though all of them are wonderful and can be eaten regularly. The recommended green and leafy vegetables include the following.

Green and Leafy Vegetables

Asparagus

Broccoli***

Brussels sprouts***

Cabbage***

Chinese cabbage*** (also known as Napa cabbage)

Collard greens***

Green peas

Dandelion greens

Endive

Escarole

Kale***

Leeks

Mustard greens***

Parsley

Scallions

Snow peas

Watercress***

I have specifically omitted dark green romaine lettuce from the list. Many people think that as long as they are eating lettuce, they are getting vegetables. That's a mistake, in my view. Dark lettuces are wonderful and healthful foods. But they do not provide the nutrition, grounding, and immune-boosting power that cooked collards, broccoli, kale, watercress, or many of the other green vegetables provide. I also urge you to eat a variety of green vegetables. Each vegetable provides its own unique array of nutrients and energetic properties. You don't want to limit yourself to salad.

Therefore, I recommend that you do not include lettuce as part of this step. You can eat dark lettuce, such as romaine, whenever you have the chance, or desire it, but think of it as an addition to the step, and not the equal of the other green vegetables. (Avoid iceberg lettuce, however; it contains virtually no nutrition.)

The Rich Benefits of Plant Foods

Plant foods are the primary source of nutrition for humans. People think of animal foods as good sources of certain nutrients, especially minerals, such as calcium. But the only reason milk products contain calcium is because cows eat plants. All minerals come from the earth. Calcium, iron, phosphorus, and other minerals make their way into the food supply because they are absorbed by plants and then the plants are eaten by animals, which in turn are eaten by humans. If cows suddenly stopped eating plants, there would be no calcium in milk. Therefore, I often say, let's eliminate the middleman—or in this case, the middle cow. Let's start going directly to the plants for our nutrition.

As for vitamins, plants are the real sources. This is especially important if you want all the vitamins, phytochemicals, and fiber your body needs.

There are 13 essential vitamins, 11 of which are found in abundance in plants. The only two vitamins that are not found in plants are vitamins D and B_{12}. Ten to twenty minutes of sunlight, three times per week, even on a cloudy day, is enough to provide your body all the vitamin D it needs to sustain health. I recommend that we get out for at least 20 minutes a day, every day—weather permitting, of course. Nutritionists recommend taking a broad-spectrum multivitamin that contains 200 to 400 micrograms international units, if there is any concern about getting adequate vitamin D. Vitamin D is fat soluble, which means we store it for a long time in our tissues. B_{12} is made from bacteria and is present in all animal foods, including fish. Like vitamin D, B_{12} is also fat soluble. The body can store an adequate supply of B_{12} for many years (some scientists believe we can store ample amounts for up to 20 years). If you eat fish or any other animal protein sources, as I recommend you do, you will obtain more than enough B_{12}. In addition, you'll be storing what your body doesn't need to be used later.

As for antioxidants and phytochemicals (both discussed in depth in Chapter 8), plants are the only real sources. The only antioxidant that animal foods provide is vitamin E, which is present in small amounts in a few animal foods. Even in the case of E, plant sources still provide more.

Plants are the only source of vitamin C and beta carotene, as well as most other antioxidants. They are also the only source of phytochemicals (most of which act as antioxidants and immune boosters), plant estrogens (which protect us against cancer), and fiber. Plant foods provide a wide spectrum of vitamins, minerals, and other health-promoting substances.

Not only are green and leafy vegetables among the most nutritious foods on earth, they are also nature's fast foods. It takes less than seven minutes to wash and steam a big bunch of collards or mustard greens, cabbage, or broccoli. Remember, eat at least three servings of green and leafy vegetables a day. Serving size doesn't matter.

Recipes for all the foods I recommend, as well as sauces and condiments to make them even more delicious, can be found in Chapter 10. Here are some quick tips for cooking vegetables.

Tips for Cooking Vegetables

The best methods to prevent loss of nutrients are steaming, boiling, and sautéing.

Steaming Steam for three to five minutes, depending on their size and consistency, over a half inch of water.

Boiling Boil for three to five minutes in enough water to cover vegetables; add a pinch of sea salt or a couple of drops of shoyu (naturally aged soy sauce; see recipes; optional). Nutrients are lost in greater quantities when vegetables are boiled in a larger volume of water over a longer period of time. Reuse vegetable broth in soups and sauces to add nutrients.

Sautéing Sauté with good-quality oils (olive oil or sesame oil are ideal). Lightly coat frying pan, add washed and cut vegetables, and sauté for about five minutes.

Oils to Use

High-quality extra virgin olive, sesame, flaxseed, and sunflower oils are my first choices for the use of oil as a dressing and for sautéing. Oil is liquid fat and can add weight to your body. A tablespoon of olive oil provides 125 calories. Therefore, I recommend you use it sparingly. Recipes typically call for about one tablespoon of olive oil per serving of salad, or as a condiment on vegetables. If your weight is not an issue, two tablespoons of olive oil per day is not a problem.

Oils to Avoid or Minimize

The following oils can be harmful because of their high saturated fat content and/or their capacity to rapidly breakdown and become rancid.

Coconut

Palm kernel

Peanut oil

Soybean oil (The most commonly used oil in processed foods. Use organic soybean oil whenever possible to avoid genetically engineered soybeans.)

Stick margarine

Salad dressings with made with hydrogenated oils, soybean oils, egg yolks, and cream.

Vinegars and Condiments

High-quality vinegar is a delicious way to add flavor to your food, without adding fat. In Chinese and other forms of traditional medicine, vinegar helps to purge and cleanse the liver.

Balsamic vinegar

Mirin

Umeboshi vinegar (A Japanese vinegar made from salted, pickled plums. Very tangy and delicious.)

Rice vinegar

Wine vinegars

The following condiments can be used to flavor your food:

Grated ginger root

Fresh horseradish

Lemon and lemon juice

Ketchup

Mustard

Pepper

Pickles

Roasted sesame seeds (*see* seeds and nuts)

Roasted sunflower seeds

Sauerkraut

Salsa

Shoyu, tamari, miso

Note: Avoid adding salt as a condiment at the table and use it only sparingly in cooking. (Usually a pinch is all that is needed.) There are a variety of healthful salt substitutes available at supermarkets and natural foods stores.

For more condiments and sauces, see Chapter 10.

Step 4

Eat at least one sweet vegetable every day.

Recommendations

- Eat at least one serving per day of sweet vegetables. These foods are delicious, filling, and satisfying. They are packed with immune-boosting and cancer-fighting fiber, antioxidants, carotenoids, and phytochemicals. They also extend the energy wave and help to heal the pancreas.
- Use the following recipe titled "sweet sensation" as a guide to preparation.
- Vary the sweet vegetables you use among those recommended in this chapter.
- Regularly include roots and other semi-sweet vegetables, listed later in the chapter.
- Report any changes in your cravings and health in your food and health diary.
- Combine Steps 1, 2, 3, and 4 of the Energy Balance program for at least one week before adopting Step 5. At this point, you will be drinking more water, eating with greater consciousness, eating at least two servings of green vegetables, and one serving of sweet vegetables each day.

Almost everyone craves sweets. Rather than depend on sugar, candy bars, and desserts, I have found that if you add more sweet flavor to your daily diet, your craving for sweet treats will be reduced.

There are certain vegetables that, when cooked, have a deep, sweet flavor. These include corn, carrots, onions, beets, winter squash, sweet potatoes, and yams. There are also less popular vegetables that provide a kind of semi-sweet taste, such as turnips, parsnips, and rutabagas.

As I explain in detail in Chapter 8, Chinese medicine has long used these vegetables to heal the pancreas and spleen. Chinese medical doctors have found that when these organs are imbalanced, or functioning below optimal levels, they create severe cravings for sweet foods, especially for sugar. In addition to the sweet vegetables I have mentioned, several other vegetables have healing properties for the pancreas and spleen. These include onions, red radishes, daikon radish, green cabbage, red cabbage, and burdock.

When these foods are eaten on a daily basis, they heal these organs and drastically reduce our need for extreme foods.

Sweet Sensation: Sweet Vegetables Cooked Together

Add one, or as many as five, sweet vegetables to a pot. Chop the hardest ones, such as the carrots and beets, into small pieces. Softer vegetables, like onions and cabbage, can be cut into larger chunks.

Remember, the ones on the bottom will get cooked more than the ones on the top. Add enough water to cover the bottom of the pot; you may want to check the water level while cooking. Add a pinch of sea salt or a few drops of shoyu or tamari to enhance flavor. Cook for approximately 20 to 30 minutes, give or take, depending on the type of vegetables, and how many you have included in the pot. This produces a delicious, satisfying, and very filling dish.

After the vegetables have been prepared, empty the ingredients into a large bowl and use the leftover cooking water as a delicious, sweet sauce.

You can also blend the ingredients together and make a creamy carrot soup or some other creative variation that can be used as a soup or sauce.

Sweet vegetables can be prepared in a matter of a few minutes. Like green vegetables, sweet vegetables can be prepared quickly by people who are busy and tired and who want to eat a healthful meal quickly, without a lot of fuss. Even sweet sensation can take less than 20 minutes. Eat these foods daily and watch your cravings diminish within a week or two.

Sweet Vegetables

Carrots

Corn

Beets

Parsnip

Onions

Squashes:

Acorn squash

Butternut squash

Delicata squash

Hubbard squash

Yellow squash

Hokkaido pumpkin

Kabocha (a name derived from the word *w*)

Pumpkin

Zucchini

Sweet potatoes

Yams

Vegetables with Healing Effects on the Pancreas

Burdock

Daikon radishes

Red radishes

Turnips

See Chapter 10 for recipes for cooking sweet vegetables.

I cannot stress enough how important it is to include sweet vegetables in your daily diet as a way of reducing your craving for sweet flavor. The more we get our sweet flavor from vegetables, the less we crave processed foods, pastries, and candy.

Step 5

Eat a protein-rich food at least once a day.

Recommendation

- Eat at least one and, if desired, as many as three protein-rich foods per day.
- During the first month, protein-rich foods include beef, pork, lamb, poultry, eggs, and fish.
- Limit the portions of red meat, poultry, and fish to 3½ ounces, about the size of a deck of cards.
- After the first month on the diet, reduce animal foods to a maximum of one a day, and preferably two to four times per week, depending on how much you desire them. Animal foods include eggs and/or lean portions of red meat, lamb, pork, poultry, and fish. By limiting animal foods, you will reduce your cravings and create a more stable state of balance.
- Substitute beans and bean products for animal foods as a primary source of protein. Use beans or bean products four to six times per week.

If you find that you still crave processed foods and sugar after reducing the meat to once a day, or four times per week, then reduce animal foods even further, preferably to once a week.

As I stated in Chapter 3, the Blood-Type diet, while excessive in its recommendations, can be a helpful guide to how much you may need animal proteins in your diet. The following guide can help you determine how much animal food you can eat, based on your blood type.

After one month on the diet, reduce animal foods as follows:

- People with type O blood can eat animal foods up to once a day, but should attempt to limit them to four days per week. That means that for three days per week, the meals will be composed entirely of vegetables, beans, grains, noodles, fruit, and other snacks and desserts.

- People with type B blood should eat animal foods between two and four times per week. That means that at least three days per week, and as many as five, the meals will be composed entirely of vegetable-quality foods, including grains, noodles, vegetables, beans, fruit, and plant-based snacks.
- People with type A blood can experiment with vegetarianism. However, if animal foods are desired, try to limit them to two or three times per week. A blood types often require the least amount of animal foods in their diets.
- For at least a month, and preferably three months, avoid all dairy products, including all forms of cow's milk, cheese, and yogurt. Milk is discussed at length in the following sections.

Finding the right amount of protein for your diet is one of the keys to your energy levels, as well as your emotional and mental well-being. People intuitively know that animal foods, such as beef, pork, and eggs, increase one's sense of personal power, self-confidence, and mental aggressiveness. That's one of the primary reasons why people don't want to give up eating red meat. They know it makes them feel strong and better able to take on the world. It's also the primary reason why the typical American diet is dominated by animal protein.

As I explain in Chapter 8, excessive amounts of animal foods increase chemical neurotransmitters in the brain called dopamine and norepinephrine, which increase our sense of alertness, aggression, personal power, and confidence. All of that is great, but when we eat too much of these foods, they can make us increasingly aggressive, anxious, fearful, controlling, and—at the extreme—even a bit paranoid.

Whole grains, many vegetables, and beans increase another chemical neurotransmitter in the brain called serotonin, which promotes a sense of well-being, optimism, relaxation, deeper and more restful sleep, and greater emotional balance. As you shift to a more plant-centered diet, you naturally become more balanced and emotionally centered. This is a wonderful experience. After years of anxiety, nervous tension, and fear, people experience greater personal security, optimism, and faith.

But just as with excess animal-food consumption, a plant-based diet has a downside, too. With time, you can become too balanced, too centered, and introverted. You may experience a lack of excitement, enthusiasm, and a certain *joie de vivre*. Plant foods can make you so balanced and centered that the outside world seems excessively tumultuous, aggressive, and overbearing. All of those feelings can be remedied by increasing the amount of animal foods you eat. The trick to eating to support your lifestyle is to achieve your own personal balance between plant and animal foods. You can do that by maintaining a plant-centered diet, while finding just the right amount of protein to suit your needs. Once you have done that, you'll experience the personal power to take on the world, but also the centeredness to remain calm and balanced in the face of the challenges life presents you.

Recently, one of my students wrote me and her classmates a letter that beautifully described the process that she went through to achieve that very balance. My student wrote this letter after I spent a few hours lecturing on the importance of protein and how it makes us feel. Here's her letter, slightly edited to maintain privacy.

> Dear Joshua,
>
> I made an incredible discovery. Changing the way I eat in the last four months has been an interesting experiment and learning experience. I learned that I can live without dairy and I like it, and that eating more greens and whole grains makes me feel more in touch with my body and who I am. However, since the New Year, I've eaten no red meat or chicken, only fish for animal protein, which I love. I've noticed that I've become more of a homebody and less energized and outgoing than I usually am. After your class this weekend, I realized that my introversion may be due to my lack of protein from red meat. So, yesterday, I had a snack of a few pieces of chicken in my leftover quinoa/tofu/kale/chard stir-fry and then I shared duck with a friend at dinner later in the evening.
>
> I am so amazed. This morning I popped out of bed at 7:30 A.M. and have been totally productive and energized and really positive. Also, my friend last night asked if she could be my client. [At my school, I train people to become teachers of the Energy Balance Diet.] She wants help with feeding herself and her husband and kids healthy foods. All I can say is, there's the Universe presenting an opportunity to me after I learned a lesson in how to take care of myself. I'm so psyched. I am really getting to know my body. Also, I love the idea of the smaller portions of meat.

Beans and Bean Products as a Source of Protein and Health

Beans and bean products are among the greatest health-promoters in the food supply. They are rich in protein, complex carbohydrates, fiber, and phytochemicals that help prevent all forms of cancer, especially breast cancer in women and prostate cancer in men. They're rich in nutrition, especially B vitamins, potassium, phosphorus, and magnesium. (The health effects of beans are fully described in Chapter 8.)

High-quality organic beans can now be purchased in jars in most natural foods stores and many supermarkets. All you have to do is reheat them and voilà, you have beans for dinner.

Among the beans I recommend are the following:

Aduki
Black beans
Black-eyed peas
Chickpeas
Kidney beans
Lima beans
Lentils
Navy beans
Pinto beans
Soybeans

I also encourage you to try soybean products, including the following:

- **Edamame.** Widely available in supermarket freezers, edamame is whole soybeans, still in their shells. They are boiled and ready to eat in less than ten minutes. Garnish lightly with salt. They are eaten by popping the soybeans out of their shells and into your mouth. Edamame is a wonderful evening snack. As with most soybean products, they are rich sources of calcium, protein, and isoflavones.
- **Miso.** A Japanese product made from aged and fermented soybeans that is converted into a paste and used as a base for soups, stews, and sauces. Miso, like other beans and bean products, contains genistein, a chemical that kills tumors by blocking blood vessels from attaching to them.

There are a wide variety of misos, including barley, rice, chickpea, and millet. In addition to providing all the benefits of the soy foods, miso provides friendly flora and digestive enzymes. For this reason, you should not boil miso—that would kill the flora and enzymes—but rather add it after a soup or stew has been fully prepared and the flame turned off.

- **Shoyu.** The name given to naturally aged and fermented soy sauce, shoyu contains friendly flora and digestive enzymes. You should purchase only the shoyus that contain no artificial colors, flavors, or stimulants to fermentation. Shoyu does contain sodium—about 18 percent, which is about half that of normal table salt. You can purchase low-sodium shoyu, which contains about 25 percent less sodium.
- **Tamari.** The liquid byproduct of miso, with all the friendly flora and digestive enzymes.
- **Tofu.** Rich in protein and calcium (about 150 mg in a 4-ounce serving), and abundant in isoflavones, including genistein. One of the richest sources of phytoestrogens, tofu can be eaten raw or cooked. It can be cut up into small squares and cooked in soups, noodle broths, and stews. It can also be baked.
- **Tempeh.** Tempeh is fermented soybeans that are compressed into a patty. In addition to providing the protein, calcium, and isoflavones of tofu, tempeh provides friendly flora that promote healthy digestion. Tempeh can be fried or boiled. It is delicious fried in sesame oil and then added to soups or noodles and broth.

Cooking Beans

Beans are so satisfying and delicious that once you get into the habit of cooking them, you'll forget all about the fact that they are so health-promoting. Whenever possible, soak your beans overnight to make them more digestible and to eliminate the phytic acid that can promote indigestion. Here are a few ways to cook beans.

- **Boiling.** The preferred method for cooking most beans. Soak beans overnight and boil them with a single stalk of kombu seaweed, which will cause the beans to be more digestible. Add a pinch of sea salt or a few drops of tamari or shoyu when the beans are 80 percent done. Boil for about 1½ to 2 hours.

- **Pressure cooking.** (Be sure that the pressure cooker regulator is clear and clean before pressure cooking beans, because beans can clog the regulator and cause problems in cooking.) Add three cups of water per cup of beans; cook with a stalk of kombu seaweed. Cover pressure cooker, lock shut, bring to pressure, as indicated by the hissing of the regulator (usually requires approximately 10 minutes), reduce heat to low, and cook for 45 minutes.
- **Baking.** Place beans in pot of water; three to four cups of water per cup of beans; add stalk of kombu seaweed, if desired, and a pinch of sea salt. Bake at 350°F for 3 to 4 hours. When beans are 80 percent done, add a variety of condiments or spices, such as raisins, miso, tamari, shoyu, others. Tempeh and tofu can be fried, baked, steamed and boiled; they can be added to soups and stews. Tofu and natto can be eaten raw.

Food has a powerful affect on us. Small amounts of animal food can change our energy levels, as well as our perspective on ourselves and the world. Here are my recommendations for finding your own protein balance. Remember, after you have been on the Energy Balance Diet for a few months, you very likely will want to experiment with your protein intake in order to find just the right amount of protein for your body and lifestyle.

Step 6

Eat at least one serving of cooked whole grains daily. Grains are the anchor in the diet. They are rich sources of nutrition, fiber, and complex carbohydrates, which makes them great sources of energy.

Recommendations

- Eat at least one serving of cooked whole grains each day.
- Choose the quick-cooking grains, especially if you are pressed for time.
- Boil your grains as the primary way to prepare them. Occasionally pressure cook.
- Use a crock pot to cook grains overnight—you'll have hot cereal waiting for you in the morning—and during the day so that you can cook rice while you are at work.

- Whole-wheat noodles and Japanese soba (made of buckwheat), udon (sifted wheat), and other high-quality noodles can be used as substitutes for whole grains. If you have a wheat allergy, substitute for wheat noodles those made from buckwheat, quinoa, potato, spinach, artichoke, or any of the other vegetable or grain noodles.
- Whole-grain bread, including sprouted grain bread, should not be considered a substitute for cooked whole grains. As I have said, the carbohydrates in bread are rapidly absorbed; bread is also rich in calories; and many people have allergic reactions to wheat.

I recommend that you start out using the quick-cooking whole grains, such as amaranth, millet, quinoa, spelt, and teff. Eat rolled oats and steel cut oats for breakfast. All of these foods can be prepared in 30 minutes or less.

Once you start incorporating these whole grains into your diet, add brown rice, sweet rice, barley, and bulgar. These grains take longer to prepare, but their nutritional value and flavor make them well worth it.

Whole-grain and high-quality pasta, including Japanese udon (made from sifted wheat), soba (made with buckwheat), and jinenjo (made from a small Japanese potato, called jinenjo), provide complex carbohydrates and nutrition, including B vitamins, iron, and fiber. (See recipes in Chapter 10.) Noodles have a relaxing effect on the body. Many people mistakenly believe that they are high in calories, but a 3½-ounce serving of pasta with tomato sauce and vegetables provides about 160 calories, a little less than the same size serving of chicken and about 60 calories less than a 3½-ounce serving of sirloin steak. The reputation for being high in calories comes not from the noodles themselves, but from what's added to them—namely cheese and meat.

Whenever possible, add green, leafy, and sweet vegetables to your pasta dishes to make pasta primavera. This is an ideal way to eat pasta, since you increase its nutritional value and make it more filling on fewer calories.

Bread should not be considered the equal of whole grains. Bread is a very drying food—the moisture needed to digest bread has to come from the body, because it's not in the food itself. This makes it difficult for many people to digest. Also, once the grain is cracked, it starts to decay. In the process, it loses nutrients and vital energy. As I pointed out in Chapter 2, bread is rapidly absorbed, just as sugar is. This creates an extreme energy wave and results in weight gain, fluctuations in energy, hypoglycemia, and mood swings. For all these reasons, bread should not be used as a substitute for whole grains.

The Value of Whole Grains

Eating cooked whole grains may be a big step. The introduction of these foods, if you don't eat them already, will have a profound affect on your physical, emotional, and psychological health and well-being.

The complex carbohydrates in whole grains are slowly burned, giving you long-lasting, enduring vitality over many hours. This extends your energy wave, as I explained in Chapter 2. Whole grains are rich in nutrition, fiber, and an amino acid called tryptophan, which increases your brain's levels of serotonin. Just as the complex carbohydrates endure for many hours, so too does the increase in brain levels of serotonin. This long-lasting presence of serotonin provides an enduring sense of well-being, optimism, and balance.

People who eat whole grains experience the kind of physical, emotional, and psychological integrity that I described earlier in this chapter. Part of this is due to the nutritional and energetic integrity of the food itself. Each grain is a matrix of whole, unified energy and nutrition. Crack the grain and it rapidly decays. Keep it whole and unprocessed and it will hold its nutritional value—and its capacity to act as a seed—for centuries. The tomb of Tutenkhamun contained rice that was placed in the tomb more than 4,000 years ago. When the king's tomb was excavated in the twentieth century, the rice could have been planted and would have grown seedlings. Its life force was still intact. You consume that whole, integrated, and unified life force when you eat whole grains.

Most people get their grains from processed foods, which have an extreme effect on the body, especially on blood sugar, insulin, and weight. Processed foods are made up of simple carbohydrates, which begin seeping into the bloodstream the minute you put them in your mouth. They cause a rapid rise in blood sugar and insulin, followed by a rapid decline in both. The net effect of processed carbohydrates is to leave us with low energy, fatigue, anxiety, and cravings for more carbohydrates and nutrition.

Cooked whole grains are another form of carbohydrate entirely. These foods are the greatest sources of enduring energy and emotional stability in the food supply.

The carbohydrates in whole cooked grain are complex, meaning that they are long chains of sugar that must be worked on by your digestive system in order to be released into the bloodstream. In addition, those complex sugars are tied up inside the grain's fiber. In order for those sugars to be absorbed and utilized by your body, they must be broken down inside your intestine and freed from the fiber. All of this takes time and effort inside your small intestine. The result is that the sugars in, say, a bowl of brown rice are extracted slowly from the grain and released into your bloodstream over a period of hours. If you eat a bowl of brown rice in the morning, the carbohydrates in that rice are going to be released into your system over several hours. That means that you're going to have energy right up to lunchtime. If you eat another whole grain at lunch, you're going to have enduring energy right up until dinner. Another grain at dinner gives you enduring and abundant energy right up to the time you go to sleep.

Because they are whole and unprocessed, grains contain all the vitamins, minerals, phytochemicals, protein, and fat that nature blessed them with. Harvard University agronomist Professor Paul C. Manglesdorf said, "A whole-grain cereal, if its food values are not destroyed by the overrefining of modern processing methods, comes closer than any other plant product to providing an adequate diet." I am not suggesting that we eat only whole grains, but when it comes to satisfying your overall needs for nutrition, grain is the most nutrient-rich food available.

Whole grains include brown rice, barley, buckwheat, millet, oats, and whole wheat.

See Chapter 10 for information on how to prepare whole grains.

Step 7

Eat health-promoting snacks. Believe it or not, eating between meals can keep insulin levels down and promote weight loss. The trick is eating healthy foods, rather than processed stuff that has no nutritional value, lots of calories, and, in many cases, lots of fat. Snacking's got a bad name because of what we snack on.

Recommendations

- Snack two to three times per day, between meals.
- Whenever you snack, choose a whole food first, such as a vegetable, a fruit, or a cooked or leftover dish from a previous meal.

- Use soups (either homemade, or high-quality store-bought soups, or dehydrated vegetable soups) as a snack. By high quality, I mean an emphasis upon organic ingredients and/or the absence of artificial ingredients in the soup.
- Make udon or soba noodles in tamari or miso broth as a snack. See recipes.
- Eat the recommended nuts and seeds as a snack or dessert.
- Avoid processed foods that are dry when eaten. (This means that you can enjoy dehydrated soups.)
- See recipes in Chapter 10 for recommended desserts.

Recommended Snacks

Here are some of the snacks I recommend:

- **Fruit.** All fruit is fine.
- **Soups.** Whenever possible, make homemade soups. However, there are many good-quality, dehydrated vegetable soups. Just add hot water for an instant, nutritious snack. Many are composed of organic ingredients. See Appendix A for recommended soups. Also, you can make a large pot of soup, refrigerate, or freeze and then serve portions whenever you want a quick snack.
- **Nuts and seeds**. Specifically walnuts, almonds, pumpkin, sunflower, and sesame (sesame seeds are most satisfying as a condiment on grain). Combine sunflower seeds and/or nuts and raisins as a snack.

  Walnuts and almonds lower cholesterol levels, especially LDL cholesterol, the type that causes heart disease. Walnuts contain selenium, a mineral antioxidant. Though walnuts, almonds, and sunflower seeds contain significant amounts of fat, most of it is polyunsaturated and monounsaturated fats. People who eat walnut and almonds tend to be more satisfied with their food and therefore tend to eat less, researchers have found. Those who eat nuts also tend to be healthier and leaner than those who do not eat nuts.
- **Noodles** in broth as a late night snack. See Chapter 10.

Sweeteners

As I have said, we all crave sweet flavor. Here are sweeteners that can satisfy your sweet tooth, without burdening your health.

Barley malt

Honey

Rice syrup

Maple syrup

Use them sparingly.

Barley malt and rice syrup are great on a rice cake when you want something sweet and mild, without a lot of calories or unhealthy oils.

Dips and Spreads

Here are a few of my favorite spreads that can provide flavor to otherwise dull rice cakes or other healthful crackers.

Apple butter—great on rice cakes

Babaganoush—a spread made from eggplant

Bean dips—pinto bean, black bean, and others

Fruit spreads—make sure they don't have added sugars

Guacamole

Hummus

Salsa

Vegetables

A pot of vegetables can be steamed or boiled in less than 10 minutes. Add one of the condiments mentioned above and you'll have a delicious and satisfying snack any time of the day. Fresh raw vegetables can be brought to work to snack on at your desk. Not only will they satisfy your need to much on something, but they'll promote your health, as well.

Any and all raw vegetables, such as celery, carrots, tomatoes, and salad.

Any cooked vegetables as a midday snack. Broccoli, with rice vinegar and olive oil, or sautéed collards, kale, or some other green can be a wonderful snack.

Prepared Snacks

There are some prepared foods that can be used as occasional fun and healthful snacks. Here are a few of my suggestions:

Low-fat trail mix—minimize because of the calories

Low-fat granola—minimize because of the calories

Popcorn, without butter—minimize because of the calories

Rice cakes—a wide variety to choose from

Desserts

See Chapter 10.

Step 8

Significantly reduce or eliminate the following foods from your diet.

- Red meat, including steak, veal, venison, lamb, and pork. If you desire these foods, limit them to 3½ ounces, as explained earlier in the chapter.
- Eggs. If eggs are desired, limit them to two per week.
- Limit poultry, both chicken and turkey, to 3½ ounces, as described earlier in the chapter.
- Avoid refined white sugar and foods that contain white sugar.
- Avoid all forms of white flour products, such as white bread, rolls, pastries, and doughnuts.
- Avoid all forms of candy that contain refined white sugar.
- Avoid soft drinks, including all cola beverages, energy drinks (such as Gatorade), and sodas.
- Eliminate coffee. Substitute various kinds of grain coffees and black and green tea to reduce caffeine withdrawal.

- Eliminate hard liquors.
- Eliminate all foods that contain artificial preservatives, colors, and flavors.
- Avoid all forms of dairy products for at least one month, and preferably three months. Follow the recommendations provided in the following section.

Restrict All Dairy Products

Dairy products are one of the greatest sources of illness and allergies in the food supply. Here are my recommendations for weaning yourself from cow's milk and milk products, and for how you can reintroduce them safely if you desire these foods.

Recommendations

- For at least one month, and preferably three months, eliminate all milk and dairy products, including whole, skim, half-and-half, and low-fat milk, cheese and yogurt.
- Note any changes in allergies, frequency of skin rashes, colds, mucous discharge, asthma attacks, arthritis, joint pain, digestive disorders, constipation, stomach upset, and immune-related problems. Many people experience the elimination of these disorders after they have stopped eating dairy products.
- You may reintroduce small amounts of high-quality yogurt, other high-quality milk products, or goat's milk cheese, if desired, after you have discovered the effects dairy foods have on you. By high-quality, I mean from milk products that are certified organic and have come from cows that have not been exposed to bovine growth hormone (rBGH).
- If you reintroduce dairy into your diet, be sure to note any changes in your health as a consequence of these foods. If any chronic symptom returns, reduce the amount of dairy foods you are eating; change the frequency with which you eat them; and change your choices. Many people cannot tolerate any cow's milk products and realize, only after they have stopped eating them, that these foods are a source of chronic illness and discomfort.
- If cow's milk products cause distress, experiment with small amounts of goat's milk cheese, or soy cheese, soy milk, and rice milk.

Greater Benefits Than You Can Imagine

After using this program to help thousands of people, I have found that it is especially effective against the following disorders:

- Heart disease.
- Overweight and obesity.
- Female-reproductive problems, including PMS, menstrual distress, fibroids, and cysts.
- Eating disorders, including anorexia, bulimia, binge eating, and severe cravings.
- Allergies.
- Addictions to sugar, chocolate, ice cream, cookies, coffee, cigarettes, and marijuana.
- Parasites, chronic fatigue, and candida albicans.
- Unusual and puzzling conditions.
- Mental confusion and stress-related disorders.
- Chronic mild to moderate depression.

Conclusion

In Chapter 10, I provide three weeks of breakfasts, lunches, dinners, snacks, and desserts for the Energy Balance Diet. I recommend that you adopt the diet as it is presented for at least a month. You may want to stick with this program even longer, especially if you want to lose weight or overcome a specific health problem. The program in its basic form can provide you with the fastest path to health.

In Chapter 8, I show you how to modify this diet to address specific health concerns.

Once you have experienced the effects of this program, and the balance that naturally flows from it, you may want to start experimenting with your daily regimen, especially with the amount of protein you consume each day. With prudent choices and conscious eating, you will soon discover your own diet—the way of eating that leads you to ever-improving health, well-being, and a deeper sense of the inner you.

chapter 5

Love: The Most Essential Food of All

Some years ago, a woman I had known for some time came to me to help her lose weight. She had been on several diets before, had lost some weight and then regained it, and then gained some more. She wanted to know if I could design a diet that would help her to lose weight, but would give her the freedom to enjoy her life, too.

"Why do you have a problem with weight?" I asked her.

"I guess I eat too much," she said. "Or maybe I eat the wrong foods."

"Why do you eat too much?" I asked.

"Because I feel a lot of tension, especially around the time of my ovulation and period. I find myself craving chocolate and sweets and stuff like that."

"Why do you think you have a lot of tension around those times?" I asked.

She looked a little embarrassed at first and then said, "I don't know. Maybe it has something to do with sex and relationships."

"Good," I said. "Have you had problems with sexual relationships?"

"Well, I was sexually abused as a child," she told me.

"Do you think that sexual abuse has anything to do with your weight?" I asked.

"Yeah, I really think it does," she said. "Many times I have thought that maybe I put on weight to protect myself from men. I mean, I want to be in a relationship, but I have to be really selective. I've often thought that if I was thin, I'd have all these guys around me and then I'd be scared and confused. I don't want that. On the other hand, I still want to be in a relationship. As long as I'm not, I eat. Maybe it's a way of loving myself—you know, to relieve the tension I feel about not being in a good relationship."

"Do you think that food is going to heal the wounds caused by sexual abuse?"

"No," she said.

"Do you think that food can replace having a good man in your life?"

"No."

"Do you think that food can replace being touched in a loving way?"

"No."

"But isn't that what you're asking food to do for you now?"

"Yeah, sort of," she said. "But I can't just walk out on the street and grab the first guy and say, 'Hey, can you be my perfect man?'"

"You're right," I said. "But maybe you could talk to someone who is really knowledgeable about sexual abuse, someone who can help you heal the wounded part of yourself. In time, you might realize that you can be in a relationship and still be in your power. You have choices as an adult that you didn't have as a child. You have the power to say yes to some things and no to others."

I recommended a support group of women who help other women deal with the issues of sexual abuse. I then described the Energy Balance plan, which I thought could help her.

As it turned out, this woman went to the women's support group and started talking about her very emotional experiences. Gradually, she started to lose weight. In fact, over the next three years, she lost all of her excess weight and kept it off. One day she came to see me and said that she had found it easier and easier to eat healthfully. She also learned that she didn't have to be excessively strict with her diet. She could relax, eat well, and enjoy herself, too.

"How did all of this happen?" I asked her.

"You know," she said, "I didn't realize it when we first talked, but I wasn't just keeping men out of my life—I was keeping a lot of good things away. I was spending my life protecting myself from the very things I needed, and using food to compensate for my empty life. I was starving and I didn't know it."

I realized that my friend was no longer repressing her past, but allowing it to surface and become fully conscious. She was doing that within her support group that provided an atmosphere of love and understanding. In essence, she was finally giving herself the love she needed to heal. Because she was no longer repressing her memories of abuse, she didn't have the tension she once had. So naturally her relationship with food changed. She didn't need food to anesthetize her pain any longer. One of the consequences—and only one—is that she lost weight. But that was not the reason she went for help. Weight loss was one of the benefits of becoming healthier at a much deeper level of her being.

"Wow, you inspire me," was all I could think to say to her. Then we gave each other a big hug.

I am constantly being reminded that all of life is food. We take in experiences and they affect us physiologically and psychologically. There is good-quality experience, which nourishes our lives, and not-so-good-quality experience, which doesn't. We are all nourished, to varying degrees, by life.

Just as life is food, each of us hungers for life. We hunger for play, fun, touch, romance, intimacy, love, achievement, money, music, art, self-expression, excitement, adventure, and spiritual nourishment. (Not necessarily in that order, of course.) Each of us has the capacity—and the hunger—for a full life. The food on your plate cannot be a substitute for a full life. And eating is not a universal solution for life's problems.

Primary Foods and Secondary Foods

I refer to the aspects of life that are nonmaterial, experiential, but nonetheless fulfilling and nourishing, as primary foods. Primary foods are what really matter in life. They are the experiences that determine whether or not your life is enjoyable, fulfilling, and worthwhile. The purpose of secondary food is to give us energy and health so that we can enjoy the rest of our lives—so that we can enjoy primary foods.

I refer to the material food that we put in our mouths and eat as secondary food. When we use secondary food as a way of alleviating, or repressing, our hunger for primary foods, the body and mind suffer. Overweight is just one of the consequences of using secondary foods as a substitute for primary foods.

Disorders such as heart disease, cancer, overweight, obesity, high blood pressure, and diabetes are epidemic today, even among young children. The reason, very simply, is because we are filling ourselves with secondary foods when we are starving for primary foods. And we don't even know it.

Sources of Joy and Satisfaction

Secondary foods don't come close to giving us the joy, meaning, and fulfillment that primary foods give us. Let me give you an example. Think back to a time when you were passionately in love. Everything was exciting. Colors were vivid. You were floating on air, gazing into each other's eyes. Your lover's touch and feelings of exhilaration were enough to sustain you. You forgot about food and were high on life.

Or remember a time when you were deeply involved in an exciting project. You believed in what you were doing and felt confident and stimulated. Time was forgotten. Suddenly, it was evening or nighttime and you realized that there wasn't enough time in the day to accomplish all you dreamed of doing. You didn't think about eating. You didn't need to eat. Sometimes people had to remind you to eat.

When you were a child, you may have had this experience almost every day. Remember when you played outside with friends? Suddenly, it was dinnertime and your mother was calling, "Time to come in and eat." And you said, "No, Mommy. I'm not hungry yet." Once you were inside, at the table, you couldn't eat fast enough so that you could get back out and play. Probably your mother had to force you to finish your meal before you went running out the door. And at the end of the day, you returned home exhausted and fell asleep, without even thinking about eating.

When you're in love, or having fun, food doesn't cross your mind. But when you're depressed, you may be turning to food all the time for comfort and solace. Chronic depression is widespread. But so, too, are frustration, anger, disappointment, sadness, and loneliness. These conditions are all crying out for primary foods—for love, self-expression, intimacy, touch, and adventure. Instead of giving ourselves those things that we need so desperately, we feed ourselves with secondary foods—chocolate, pastries, coffee, alcohol, and drugs. You can eat all the food in the world and you will still be hungry for the things you really need.

Half-Truths That Lead Us Astray

Everybody knows that we use food as a substitute for love, or to relieve our boredom, or to entertain us, or to make us feel better about our lives, or to put our bodies, minds, and emotions to sleep. Unfortunately, our society doesn't talk about these things enough. Also, you'll never read about your need for primary foods in a diet book, even though your hunger for primary foods may be the reason you are overeating, or filling your stomach with junk food. Instead of straight talk, diet books often give us half-truths and an old sales pitch, which is that if you eat certain foods, you'll lose weight.

No one mentions the fact that you may need the unapproved foods to manage your emotions and your life. Most of us need secondary foods to keep us experiencing our intense and unsatisfied hunger for primary foods.

In addition to the immediate demands on our lives, we also have associations with food that arose when we were children. Maybe our mothers and fathers used candy or ice cream to make us feel loved. Or maybe there were special foods that our mothers made when we felt weak, or sick, or tired, or alone. We take those associations into adulthood and repeat those same patterns when we feel unloved, or tired, or sick, or alone. Thus, we often find ourselves comforting ourselves with foods that our parents offered us, even though we may consciously know that such foods are damaging our health.

Here are a few more reasons why we medicate ourselves with food every day. When people are tired, they rarely rest. Instead, they drink coffee to give themselves energy. Coffee has no energy in it. What it does have is a powerful drug called caffeine, which will draw upon your own energy reserves and force your body to work. You're still tired. You still need rest, and sleep, and relaxation. The only difference is that you used a drug to force your body to work when it was tired.

People feel tense, or angry, or depressed when they go to work, or when they come home from work, or both. They can't bear those feelings. But rather than confront them directly and change their lives so that they can be happier, they eat ice cream, or sugary snacks, or chips, or drink beer. Those feelings of anger, tension, and depression are still inside you. They're building and eventually they'll blow up. But for the time being, you use food to give yourself some temporary relief.

I'm all for relief. We all need downtime. But we also need hugs, kisses, warmth, pats on the back, massage, meditation, freedom, tears, hot baths, nature, trees, and close friends to play with. We all need to feel seen and appreciated at our work and in our primary relationships. When we live without these things, distress, inner conflict, and emotional pain arises. It's what we do with that pain that gets us into trouble. Most of us stuff it down with éclairs and doughnuts and steaks and alcohol until something in the body finally blows up.

Here's an interesting experiment: Fill up your life with primary foods; get lots of hugs, and kisses, and massage, and walks in nature. Don't think about changing your diet *per se*, but just become conscious of your food choices. Now watch your diet change naturally. And see how fast you lose weight.

Wheel of Life Exercise

Sometimes it's difficult to know which primary foods we should be concentrating on. The recommendations that follow will help you do that, but first try the little exercise provided here, which is called "The Wheel of Life." You can do it in less than five minutes. You'll be amazed at how much it reveals about the imbalances in your life. The exercise was created by Cherie Sohnen-Moe and taken from her book *Business Mastery: A Guide for Creating a Fulfilling, Thriving Business and Keeping It Successful* (Sohnen-Moe Associates, Inc., April 1999). It can help you see where your imbalances lie, and where in your life you may want to spend more of your time and energy.

The Wheel of Life illustration has 10 spokes; each one stands for a particular part of your life. Place a dot on the spoke to designate how much time, energy, and personal involvement you may be giving to that part of your life. The center of the circle is zero. A dot at the outer part of the wheel indicates very deep involvement and a strong commitment of time and energy. Once you have done that, connect the dots and see what your wheel of life looks like. This exercise can give you a good idea of where the imbalances are located, and where you may want to place more time and energy to enjoy a more satisfying life.

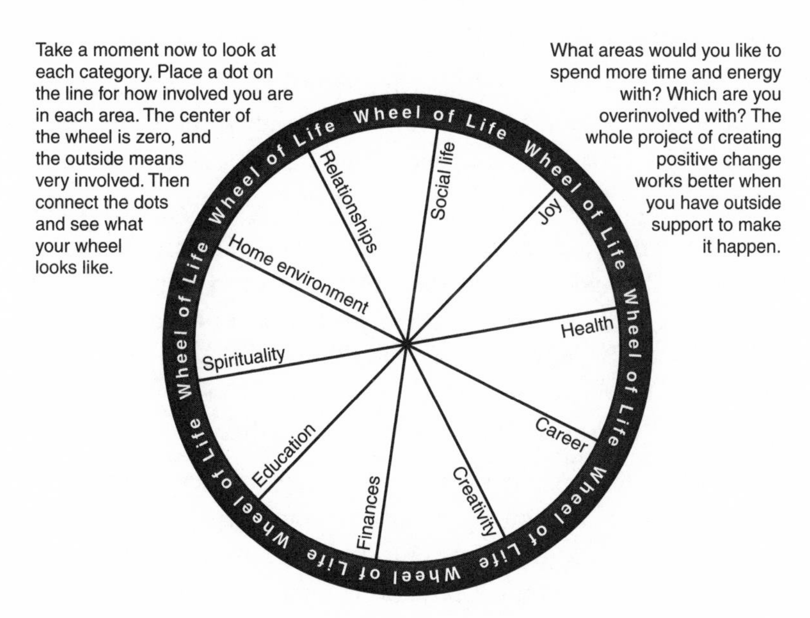

Form a New Relationship with Your Body

Your relationship with your body is like any other important relationship in your life: It requires love, attention, time, touch, and communication. Sometimes you've got to check in with your body and say, "Hi sweetheart, what's going on? What are you feeling? Am I treating you well? Do you need more attention?"

Here are several ways to address and heal deeper wounds, as well as nourish some of your primary food needs: exercise, resistance training, massage, writing, support groups, prayer.

Exercise for Body, Mind, and Heart

After eating a healthful diet, the most important behavior for changing your relationship with your body is exercise. It can boost your physical and mental health, your outlook on life, and your willpower. People who exercise regularly naturally do all kinds of other behaviors that make their lives fuller, richer, and healthier. Exercise can make you look younger and more attractive, and transform the way you see life. And the remarkable thing is that you don't have to do very much exercise to enjoy all of these benefits.

Warning: Don't Kill Yourself

Consult your doctor and get a thorough physical before beginning an exercise program. Ask your doctor for his or her advice on which exercises are best suited for your condition, and which ones you should avoid.

Do not exercise vigorously if your blood cholesterol is high, or if you have high blood pressure, unless your doctor gives you permission. A high cholesterol level is a good indicator of atherosclerotic plaque in the coronary arteries. High blood pressure is a major risk factor for heart attack and stroke. If you have any risk factors for heart disease—high cholesterol, high blood pressure, overweight, a history of smoking, or any family history of heart disease—consult your physician for his or her advice.

When first beginning a new exercise program, it's better to start out slowly. Walk, don't run. Many people who are out of shape start running in the hope of getting back into shape quickly. These people often subscribe to the "no pain, no gain" philosophy. Don't tempt fate. You're liable to get a lot more pain than you bargained for. Begin your program slowly and cautiously. Allow your body to develop fitness first. Exercise becomes more pleasurable, and addictive, when you give your body a chance to get into shape.

If you have not exercised in some time, do not participate in competitive sports. It's too easy to lose yourself in the heat of competition and either create an injury or worse.

The best exercise is simply a daily walk of 30 minutes or more. This alone will bring you all the benefits of exercise, even as it protects you from excess stress on your cardiovascular system. Whatever exercise you start out doing, purchase the appropriate clothing and shoes. The latter piece of equipment is the most important, especially if you start out walking. Buy shoes that fit you comfortably and provide plenty of support and cushion.

Remember to stretch for about 10 minutes before exercising in order to warm up, and then do another 10 minutes of stretching when you conclude your walk or aerobic workout in order to cool down.

Exercise and Live—Longer!

Here's a bottom-line benefit from exercise: If you exercise, even a little, the chances are very good that you'll live longer and you'll be healthier during the years you do live.

That conclusion was established by several studies, but the one that stands out the most was done at the Cooper Clinic in Dallas, Texas. Researchers divided a group of 13,000 men and women into five groups, according to their fitness levels and the frequency that they exercised. The scientists actually tested the people in the study to determine their level of fitness, so it was very clear what the level of fitness was among these people.

The five groups ranged in degrees of fitness. The first group was composed of the proverbial couch potatoes, people who were sedentary. The second was made up of people who walked 30 minutes a session, three or four times a week. The third, people who exercised three or four times a week. The fourth, athletes who worked out intensively four or five times a week. The fifth, marathon runners. All the people who participated in the study were followed by the scientists for eight years.

Not surprisingly, the researchers found that those who fell into the sedentary category died the soonest. That didn't surprise anyone, since scientists have long known that lack of exercise leads to an array of other illnesses. What did surprise them, however, was that the greatest difference in health between any of the five categories was between the couch potatoes and group 2, those who walked three or four times a week for only 30 minutes at a time. The jump between the sedentary people of group one and the walkers of group 2 represented the greatest differences in health, the scientists found. In fact, the difference between the walkers and the athletes who exercised much more was small in comparison to the difference between the walkers and the sedentary types.

The study, which was published in the *Journal of the American Medical Association*, found that that little 30-minute walk cut the second group's chances of having a heart attack or cancer in half. The other three groups enjoyed only slightly better health benefits than those of group 2.

Cancer rates differed dramatically among the five groups. The men who didn't exercise had four times more incidence of cancer than those who were physically fit. Women who were sedentary had 16 times the cancer rate than those who were physically active had.

Other studies have found similar results. *The New England Journal of Medicine* published a study (February 25, 1993) that found that men between the ages of 45 and 54 who take up a vigorous sport live 10 months longer, on average, than those who remain sedentary.

Exercise Makes Eating Well Easier

Exercise burns carbohydrates, the body's preferred fuel source. That creates a carbohydrate deficit, which in turn creates a craving for complex carbs found in plant foods. *The American Journal of Clinical Nutrition* (1994; 59:728S–734S) reported a study that found that people who exercise regularly desire and eat more complex carbohydrates from whole grains, vegetables, and fruits.

People who exercise regularly also eat less fat than those who do not exercise, according to a study published in *The American Journal of Public Health* (1995; 85:240–244). That only stands to reason, of course. After you exercise, your whole body feels cleaner and healthier. You just don't want to put a fat-laden burger or some greasy fries in your body after you've done so much to clean your proverbial house.

Exercise for Your Mind

What most people do not realize is that exercise dramatically improves your willpower and clarifies your goals and intentions for your life. People who exercise regularly experience greater confidence and a stronger sense of well-being. Part of this is due to the changes in brain chemistry caused by exercise. After 10 to 20 minutes of exercise—including walking—the brain increases production of endorphins, which are opium-like compounds that create feelings of euphoria, well-being, and optimism.

Such feelings are not as fleeting as many believe. Exercise has been shown to create improved and long-lasting mental health. People who have suffered from depression and other psychological disorders have experienced profound transformations as a consequence of regular exercise, according to a study published in *Postgraduate Medicine* (July 1990). Researchers found that regular exercise resulted in reduced stress and anxiety, and fewer bouts of depression. People who exercise also have greater self-esteem. They indulge far less frequently in self-criticism and have a greater capacity to deal with stress. They felt they were healthier and were much more optimistic about their futures than those who did not exercise. Aerobic exercise, including walking, had the most profound effects on mental health, as opposed to resistance or weight training.

How Much Is Enough?

Thirty minutes of "moderately intense" physical activity four to five times per week is enough to produce these and many other benefits, according to the American College of Sports Medicine (ACSM), the leading body of exercise experts. The ACSM has also found that exercise is cumulative, so that three 10-minute walks a day produce the same results as one 30-minute walk per day.

Moderate exercise has been shown again and again to produce remarkable health benefits. The *Journal of the American Medical Association* (December 18, 1991) reported a study that showed that a leisurely walk at three miles per hour can significantly reduce a woman's risk of heart attack. One of the ways it does this, scientists have found, is by increasing HDL levels, the good cholesterol that protects you from heart attack and stroke.

But even less than 30 minutes, four or five times a week, can produce a benefit. Women who walked for just an hour a week had half the risk of suffering a heart attack as women who did not walk at all, according to Harvard University's Women's Health Study. I'm not recommending less, mind you. I'm just saying get out there and enjoy yourself. The benefits will last a lifetime.

Resistance Training for Strong Muscles and Bones

What we used to call weight lifting is no longer for body builders. In fact, if you're a small woman, with small bones, you're a perfect candidate to get out there and do regular resistance training, or weight lifting. The reason: Weight training dramatically strengthens bones and muscles, and raises HDL levels, which in turn protects the heart. Resistance breaks down muscle and bone. That causes the body to rebuild both in thicker, stronger, and more flexible form. Muscle pulls on bone, which stresses bones and makes them stronger, as well. Muscle is dense and active tissue. It burns fat and excess calories, even while you are resting. The more muscle you have, the more your body is cushioned against possible falls, which can prevent bones from breaking. Bone fractures, especially those in the hips, are the reason people die of osteoporosis.

Resistance training can be done two or three times per week. After a resistance-training session, your muscles need time to heal and rebuild. Don't do resistance training two days in a row. Give yourself at least a day, and preferably two, to let your muscles rebuild themselves. Resistance training can be comprised of such simple exercises as sit-ups, knee bends, and push-ups. But the best thing to do when you take up a weight program is to join a gym and allow an exercise physiologist or personal trainer to guide you in the use of weight and resistance-training machines.

Here are some suggestions to help you get regular exercise.

- **Purchase a treadmill.** Choose one with a cushioned walking surface and use it every day while listening to music, meditative or self-improvement audiotapes, or simply in front of the television. You can do the same with a stationary bicycle.
- **Take up a sport that you can do for the rest of your life.** Golf, tennis, racquetball, swimming—these are just a few of the activities that you can do to improve your fitness, health, and weight, while you're having fun.
- **Join the Y or a local health club.** They've got the equipment, the pool, the experts, and the trainers you need to get an ideal workout at your local YMCA, YWCA, or health club.
- **Take yoga classes.** The effects of yoga on health can be truly remarkable. Not only does yoga make us more flexible, and thus significantly improve range of motion, it also can produce a deep sense of peace, emotional transformation, and balance. Yoga is physical meditation. It puts you in touch with the essence of your physical, psychological, and spiritual life. And it can have a dramatic effect on your health.

- **Dance is exercise.** People don't often think of dancing when they think of aerobic exercise, but try tango, salsa, swing, African dance, or simply ballroom and you can work up quite a sweat. But even if you don't sweat, you still can exercise your muscles (as well as your heart) dancing in the arms of someone you love. Do it for the joy, if not for your health.
- **Join a hiking club.** Most recreational departments offer organized walks and hiking clubs. It's a great way to get a workout while you meet new people and form new friendships. Many hiking clubs prepare for greater treks by getting you in shape. Consult your town's recreational department or your Yellow Pages for where you can participate in group walks.

  Of course, you can pick any park, woods, or natural environment for your own nature walk.
- **Karate chop your way to good health.** Martial arts classes are everywhere these days. Karate, Tai Chi Chuan, Judo, Tae Kwon Do, and kickboxing are available in virtually every city and most small towns throughout the United States. Most dojos, or places of instruction, offer both private and group lessons at extremely affordable rates. They are designed for people at all levels of athletic ability and coordination.

Get a Regular Massage

We need hugs and touching on a regular basis. Find someone with a very healing touch who is an expert in some form of therapeutic massage, such as shiatsu, acupressure, deep tissue, Swedish, Reiki, Alexander technique, or Jin Shin Jyutsu. Let that person work on your body to release deep levels of tension and help you heal old wounds.

Once a friend of mine hurt his back and was bedridden for three months. When he finally found an acupressure massage therapist who could help him, the acupressurist told him that he was going to put his spine back into a place that it hadn't been since my friend was in puberty. The therapist said something that was very interesting.

"By working on your back and other places on your body, I'm going to help you reconnect your consciousness with parts of your body that your mind has been out of touch with for many years. You've been putting your tension in these places for decades," the acupressurist said. "Those places have gotten so tight and stagnant that you're no longer circulating blood and lymph adequately to those tissues. The tissues are numb and asleep. You literally stopped feeling how much tension and distress those parts of your body were in. We're going to gently and tenderly wake them up again and bring them back to life."

All of us have been chronically storing tension and distress in certain parts of our bodies—the lower back, the shoulders, pelvis, and reproductive organs, just to name a few common areas where people are injured and out of touch with themselves. Someone who is very skilled and who has a healing touch can release the tension in these parts of the body. He or she will restore circulation of blood and lymph and electromagnetic energy to parts of the body that are now deprived of these elixirs of life. You will be surprised how your body and mind react to such healing. Old memories may become conscious again. Examine them, learn more about yourself, and let go of the pain that has long been associated with those events.

As I often say, "The issues are in the tissues!" So many people who have been physically abused have great difficulty being touched. They still hold the psychic wounds in their muscles, connective tissues, and skin. A massage therapist with a clear intention and a healing touch can release the tension and the painful memories from parts of the body. In the process, our bodies and our minds start the healing process—and very often, it occurs without anyone saying a single word.

Do the Pennebaker Exercise

James W. Pennebaker, Ph.D., a professor of psychology at Southern Methodist University in Texas, struggled for a long time with depression. After seeing therapists and experiencing no relief, he began writing about his emotional travails. In the course of writing about his life, and especially about his own traumas, he found himself making remarkable discoveries about himself and soon found that he was no longer depressed. Not only wasn't he depressed, but the writing seemed to trigger a kind of rebirth in him. Amazed, he decided to conduct an experiment with his students. Pennebaker devised a study in which people were to write for 20 minutes a day, for 4 consecutive days, about their most traumatic and shame-filled experience. Ideally, the students should write about events that they had never shared with anyone before, nor written about.

Pennebaker conducted immunological tests on the student participants and noted repeatedly that those who wrote about their traumatic events had higher immune reactions than his similarly matched control group, who did no writing.

Not only did the writers have stronger immune systems, but they had fewer trips to the health clinic than their matched controls. Pennebaker noted that the immune systems of those whom he termed "high disclosers," meaning those who wrote about things that were deeply personal and intimate, had the greatest immune reactions of all.

Pennebaker developed a theory as to why this was the case. He stated that repression, or psychological inhibition, which is the mechanism by which we keep things secret, even from ourselves, requires a certain degree of psychic and physical energy. As he puts it, inhibition is a demanding form of work, especially when a very painful trauma must be kept secret. Physical symptoms frequently occur from such inhibition, such as elevations in blood pressure, heart rate, breathing, skin temperature, and perspiration levels.

Pennebaker consulted experts from the FBI who conduct polygraph tests on suspected criminals and found that similar physical conditions occur among people who try to lie on lie detector tests. Once the criminal confesses, however, he becomes remarkably relaxed, and all his physical symptoms related to inhibition disappear.

The release that accompanies confession, therefore, occurs on psychological and physical levels. Pennebaker also found that when criminals confess during polygraph tests, they often bond to their confessors. Many confessed criminals later send Christmas cards and letters to the FBI agents, thanking them for their help.

Pennebaker found that confession releases an enormous amount of energy that has been used to repress painful memories and their related emotions. Once the person starts to bring up those memories and emotions, the inhibiting forces are released and psychic and physical equilibrium are restored. The result is greater feelings of peace and better physical health.

Those same events occur with the writing method, or what has come to be known as the "Pennebaker method." During the first two days of writing people experience negative emotions, such as anger, sadness, anxiety, and grief. On the third or fourth day, however, they experience feelings of relief, insight, and resolution, suggesting that they have released the inhibiting energy and integrated the traumatic events into their consciousness.

Pennebaker points out that one does not necessarily have to write about the event. Confessing it to someone else will have the same effect.

The rules for writing such confessions are simple enough:

1. Write for 20 minutes, for 4 consecutive days.
2. Write continuously about the most upsetting experience or trauma of your entire life.
3. Don't worry about grammar, spelling, or structure of the piece.
4. Write your deepest thoughts and emotions regarding the experience. Include all the details you remember and insights into the events.
5. Once you have written your deepest thoughts, you have the option of showing them to someone you trust—as Pennebaker's students did with him—or keep the journal private, or burn the book and symbolically let those experiences go up in smoke. Do whatever feels right to you. But before you do anything, consider the decision carefully. Do not share your innermost thoughts with someone who may not treat such experiences appropriately. The idea is to be heard by someone who can reflect back to you love and understanding. This reinforces your own capacity to love and understand yourself.

Do the exercise. Confine your writing to a single event. Report every single detail of the experience and every single emotion that you had during it. Write for at least 20 minutes a day—longer if you can—and see what effects, if any, the exercise has on you.

In addition, write regularly in a journal or diary. Express your feelings and write down the important events of your life. Studies have shown that people who write regularly in a diary experience significant improvements in psychological health.

Join a Support Group

There are support groups for virtually every problem humans face. Find one that's right for you. Share your pain and start the healing process. Sharing your experiences allows you to release feelings that have been trapped and repressed for many years. It also gives you the chance to discover new insights into old experiences. That alone can give you an entirely new and healing perspective on painful events.

Eat Less, Pray More

Meditation and prayer are ways of receiving primary foods from the Great Spirit, however you perceive it. By quieting the mind, meditation creates a kind of emptiness, a vacuum so to speak, that enables us to receive the more gentle energies that are always present and flowing to us, but are perhaps too subtle for us to perceive. It's the same with our stomachs. When we fill ourselves with secondary foods, we are less able to receive primary foods, including the very refined spiritual vibrations. That is why every spiritual tradition encourages fasting, or to have times during the year when you reduce your intake of secondary foods. This allows us to be more empty, more open, and more receptive to the most important of all primary foods, God's love.

Janine's Story

I first met Janine in 1996, when she was a successful actress, having appeared in several plays on Broadway and other New York stages. She had just returned from nine months on the road as part of a touring company for a Broadway hit. Whenever Janine traveled she had two suitcases, one for her clothes and another for her food. As she put it, "I was a compulsive eater. In my suitcase, I'd have pretzels and mustard, canned soups, cereals, baked flour products, cookies, water, soda—you name it. I wasn't quite bulimic, but I did binge eat from time to time and I even tried to make myself throw up a few times, but I couldn't do it. I thought about getting into that kind of behavior, though."

From time to time, Janine would fast in order to lose weight. "I'd get really skinny on some crazy juice fast," she recalled. "But it wasn't good. I felt terrible whenever I did that."

Janine also went on cleansing diets and supplement programs. "I would combine fasting and a supplement program," Janine said. "I'd be taking all of these pills, along with juices or fruit. And I would do that for four months. At the beginning of these cleanses, my period would stop. That should have told me something right away, because a woman's cycle reflects the overall balance and health of her body. You don't mess with that. When you put the body into a crisis situation, the period stops coming and that should be a sign that something is wrong. But I was so out of touch with what my body needed that I didn't recognize that."

Janine had numerous other health problems. Her menstrual periods were highly irregular; often they came every six weeks. She had terrible PMS, cramping, headaches, and emotional swings, everything from depression to acute anxiety. She had an array of digestive disorders, as well, including acid reflux, heartburn, and burning stomach.

Janine saw a massage therapist in New York and confessed her food problem to him. He recommended that she see me.

"One of the first things Joshua told me was to take a small towel into the bathtub with me, to wet the towel, and then give my body a body scrub with the hot, wet towel. That was so invigorating. It also made my body relax and feel good. I went to bed and slept much deeper after that. (More on the benefits of the hot towel scrub in Chapter 6.)

"Another thing he told me was to get more sleep. 'You've got to get horizontal by 10 P.M.,' I remember him telling me. So I did. And that little bit of extra rest made a big difference in my tension levels," Janine said. Janine's a dancer, so she didn't need an exercise program; her work is all exercise. What she needed most was rest, body care, and healing foods.

I placed Janine on the Energy Balance Diet, emphasizing lots of cooked green vegetables, sweet vegetables, grains, soups, and fruit. She also ate a variety of low-fat animal products, and some mild desserts. Within a month, the diet had balanced her condition and eliminated her intense cravings.

"It really wasn't long before I felt what the food was doing to me," Janine recalled. "I knew within the first couple of weeks that this was working. I could feel my body stabilizing. I felt more centered, more relaxed, yet I had good energy. It took about three months, but my periods came back and normalized. All the PMS symptoms vanished. A lot of that came from giving up dairy foods, I realized later. I didn't have headaches anymore, nor did I have digestive problems. My digestion and intestines functioned normally, and what a gift that was!

"My emotional life changed dramatically, as well," Janine said. "I guess the whole program gave me this feeling of being strong, yet balanced emotionally. Something opened up inside of me. I became more optimistic and sensitive at the same time. I became aware of my body in a whole new way and it felt as if I was living in my body for the first time in a long time. I started to know what my body needed. And I came to trust it."

Janine's experience is not unlike so many hundreds of others who have adopted this program. What many people discover is that healing is more than the elimination of symptoms, but the recovery of one's entire self—the body, mind, and spirit.

chapter 6
The Energy Balance Solution

A few years ago, a 28-year-old woman came to see me with severe premenstrual syndrome (PMS), heavy menstrual bleeding and clotting. She developed fibroid cysts in her uterus and suffered from fibrocystic breast disease. Her PMS symptoms included cramping, headaches, fatigue, and heavy bleeding. These symptoms were so bad that she was forced to stay in bed a couple days every month, which was wreaking havoc in her job and social life. That was not her chief concern, however. Her breasts and uterus were developing fibroid cysts and she was terrified that if things continued as they were going, she would develop some form of cancer, either in her breasts or reproductive organs. Making matters worse was the fact that her aunt had died of breast cancer. She was certain that because breast cancer was in her family, she stood a very good chance of suffering the same fate as her aunt.

This woman had been to several doctors and been on numerous medications, all without positive results. In an effort to regulate her hormones, one physician placed her on birth control pills, which apparently had made her PMS even worse. When my client read that long-term estrogen use was associated with a higher risk of breast cancer, she stopped taking the pill. Anything that increased her risk, which was already too high for her to cope with, was too much for her to bear.

The first thing that I recognized when I met this woman was her high degree of fear. I was certain that the fear was creating chaos in her life and in her way of eating. If left to her own devices, I was certain that the fear and chaotic diet would make her PMS and fibroids even worse. It was possible that her fear could even lead to the very fate that terrified her.

"Do you believe that you have the power to create good health?" I asked her.

"I don't know," she said. "So far, nothing I have done seems to have had any positive effects. My PMS is as bad as ever, maybe worse, and these cysts in my breasts and uterus really scare me."

"Okay," I said. "What if someone could give you a pill that would magically eliminate the PMS and make your cysts go away? How would you feel about that?"

"Well, I'd be happy," she said.

"Yes, and what else?"

"I wouldn't be afraid anymore?"

"Okay," I said. "And what would you have learned?"

"Nothing, I guess, but I don't care about learning anything. I just don't want to get cancer."

"Right, you'd learn nothing. Which means that you wouldn't know how to avoid getting cancer. Is it possible that if you don't know how to avoid getting sick, you might do all the wrong things over the next ten years and actually make yourself sick?"

"Yes," she said. "It's possible. In fact, I worry about exactly that happening."

"Of course. We all worry that the things we fear the most will come to pass," I said. "So maybe two things have to happen: You have to find a way to eliminate the menstrual problems and the cysts, but also to learn how to develop your health so that you can protect yourself from the thing you fear the most."

"Yes," she said. "I suppose that's what I have to do."

"That's why you're here today," I said. "Not to have someone give you something that magically makes your problems go away. You already know that that can't be done. What you're here to learn is how you can make these problems go away. And also avoid the fate that your aunt suffered. Right?"

"Yes," she said. "That's right."

"Okay. Here's what you can do."

I then explained that if she increased fibrous foods, eliminated dairy products, and reduced high-fat foods, her estrogen levels would fall significantly. That would reduce the PMS dramatically. Very soon, her symptoms would become so mild that she could have her period and still go to work. In addition, the lower estrogen levels might very well cause her cysts to shrink. Eventually, they could go away entirely. By lowering her estrogen, we could also dramatically reduce her risk of breast and uterine cancer.

"Diet can do all that?" she asked me.

"Your current diet is one of the main reasons you have PMS and cysts in the first place," I said. "If you change your diet, all the symptoms can be improved."

Three months later, her symptoms of PMS had all but vanished. They were so mild, in fact, that she needed no medication to treat her symptoms and was able to able to go to work every day of the month. She considered this improvement a miracle. A year later, she told me that her doctor had informed her that the cysts had shrunk. Her PMS had subsided even more. Her symptoms were negligible. And when they arose, she always knew what she had eaten during the previous weeks to make the symptoms appear again. Even then, the symptoms were mild, however. She was elated, to say the least.

But something even more fundamental had happened. She was empowered by what she had done and she was no longer in the state of fear that she had been when we first met. The reason for this transformation? She had learned how to control her health. That was the greatest gift she could have given herself, at least at this stage in her life.

Sicker Than Ever and in Denial

In certain respects, this woman's evolution describes the path Americans have yet to travel. Right now, people just want a pill to make their symptoms go away. There are plenty of over-the-counter medications and doctor-prescribed drugs, so people are willing to use both to eliminate problems typical of most Americans today.

At the same time, people don't know how to lose weight, regain their health, or protect themselves from disease. This makes them all the more dependent upon doctors for pills and surgery. At this point, most Americans believe that drugs and surgery are the only answers.

What people don't realize is that the American diet and the pharmaceutical industries are working at opposite ends of an addiction cycle. The diet causes the problems and the pharmaceutical industry creates drugs to treat the problems. As long as we continue to eat the standard American diet, our health problems will not only endure, but get worse. That means that the medication has to get stronger. Eventually, the medication won't work at all. At that point, the condition may be untreatable or in need of surgery.

The path most Americans are currently on is essentially the path of self-destruction. People are poisoning themselves anywhere from four to nine times a day. Day in and day out, they put excessive amounts of fat, protein, dairy foods, sugar, and processed foods into their mouths. Most people get little or no exercise. On the whole, people take better care of their dogs than they do of their own bodies.

Remarkably, most people are in denial about all of this. What helps us stay that way is that most everyone else is eating this way and taking medication, too, so it must be normal, right?

A very good case can be made that human beings are sicker than we have ever been in our long existence. The only difference now is that we have developed an enormous pharmaceutical industry that offers us pills and potions that quiet our symptoms and allow us to go on abusing ourselves until we are afflicted with an illness that medicine cannot control.

As long as we don't deal with the causes of disease, the symptoms get worse. Today, we are sicker and fatter than we've ever been. The rates of heart disease, cancer, overweight, diabetes, high blood pressure, osteoporosis, arthritis, asthma, Alzheimer's disease, and Parkinson's are many times higher than they were 100 years ago. At the same time, we are forced to consult many more doctors, undergo many more tests, take many more drugs, and suffer through many more operations than we have ever experienced before.

Do you realize that in 1900, the major causes of death were pneumonia and influenza? Heart disease was number four, and cancer was number eight. During the twentieth century, Americans ate more and more meat, chicken, dairy products, and processed foods. Today, heart disease is number one (it kills about a million people a year—unheard of in your grandmother's day), and cancer is number two (killing nearly half a million annually). At the rate we are going, cancer will overtake heart disease within the next 25 to 35 years and become the number-one killer of humans.

In 1960, a woman's chances of contracting breast cancer were about 1 in 20. In the 1970s, her chances jumped to 1 in 14. In the late 1990s, it jumped to 1 in 8. Similar rises have occurred in cancers of the prostate, colon, and lung. Most American adults are overweight—60 percent, according to the U.S. Surgeon General. At no time in history has a country's population, especially one as big as the United States', been this overweight. When you're overweight, you also suffer a much greater chance of getting heart disease, diabetes, cancer, high blood pressure, and a lot of other illnesses. There is an epidemic of overweight, obesity, and diabetes.

You're probably saying to yourself, "Yeah, but we live longer than our grandparents did." Not by much.

Average life expectancy is arrived at in the same way that an average for a series of tests are calculated. Average life expectancy is found by averaging the ages of people who die at birth with those who die in old age. That means that if the number of infants who die goes down, the overall average life span goes up (it's like dropping your lowest test score), even though no one is actually living longer. Infant mortality is going down in the United States, which is one reason why the average life expectancy is going up.

In terms of real life extension, there was only been a small increase in life expectancy during the twentieth century—six years to be exact, according to John A. McDougall, M.D., in his book *The McDougall Plan* (New Win Publishers, 1983).

The truth is that a woman who reaches the age of 45 has about the same chance of living to be 80, no matter where she lives—the United States, Spain, Brazil, or China, according to *The American Medical Association's Encyclopedia of Medicine*. In fact, people who live in some Third World nations who reach the age of 40 have a better chance of living to 80 than Americans do.

The real question we have to ask ourselves is this: What is the quality of the last 20 years of life? High-tech medicine may be able to keep us alive six years longer than it could a hundred years ago, but it is often through the use of very powerful drugs and operations, all of which have debilitating side effects. The chances are very good that during your senior years you will be dependent upon doctors and their medical arsenal to stay alive. It doesn't have to be that way.

You were designed to be healthy. A tribe of native peoples living in Botswana, called the !Kung, have the same percentage of people older than 60 years living in their society as modern America does. But the !Kung never see a doctor their entire lives. Humans were designed by nature to be independent of medical intervention, except in the case of accidents or communicable diseases, which are emergency conditions. Normal life was designed to be experienced in health.

Given the right circumstances, the human body can overcome most disorders. Part of the formula for reestablishing health is to stop poisoning the body. Take away the toxins and the body works much better. Add foods and a little exercise that nourish the body and support its healing functions and the body starts making miracles. Could it be that simple? Yes, it's that simple.

The Simple Formula for Health

We can regain our natural state of health if we give the body what it needs to restore health. Here's a simple formula for regaining your health.

1. Eat a plant-based diet, based on the four food groups I listed in Chapter 4. For shorthand purposes, I will call this diet the healthy eight. The healthy eight are as follows:

   Two servings of green or leafy vegetables per day.

   One serving of sweet vegetables per day.

   One root or round vegetable, such as carrots or onions per day.

   One serving of beans or low-fat animal foods per day. The preference for regular consumption is fish, but occasional 3-ounce portions of the white meat of chicken and lean red meats are also permissible.

   Two servings of whole grains per day.

   One serving of fruit per day.

2. Stop eating all cow's milk and milk products for at least three months.
3. Drastically reduce red meat, pork, chicken, and eggs.
4. Stop eating processed foods.
5. Walk every day or do some other form of exercise that you enjoy.
6. Scrub your body with a wet, hot towel, either in the shower, or when you get out. Do this for 10 minutes, 2 or 3 times per week. It will promote better circulation and elimination of waste products through the skin. (More about the hot towel scrub and other body care methods in the following sections.)

After a few weeks of following this advice, you'll see that this simple regimen will take care of most of your health problems. Virtually every physical health problem you have can be significantly improved by doing these five simple steps. You will very likely see improvement in two weeks. Chances are excellent that you will experience some kind of metamorphosis if you follow this advice for three months.

In this chapter, I want to explain how certain serious health problems arise and how you can protect yourself from them, or overcome them entirely, by following a simple healing program.

Lose Weight Quickly, Easily, and Without Being Hungry

If you are overweight now but lose weight, you will see a dramatic improvement in your overall health. You will look better, feel better, and reduce most, if not all, of your risk factors for major disease, including heart disease, cancer, adult-onset diabetes, and high blood pressure.

Here's a simple formula for rapid and healthful weight loss:

- Eat the healthy eight every day. (If you are eating a plant-based diet already, choose the grains I list here and reduce the amount of grain you eat to one or two small servings per day.)
- Eliminate all processed foods, including bread, rolls, pastries, and anything with flour, artificial colors, and artificial flavors. You'll lose weight very fast if you do this.
- Eliminate dairy foods. Dairy foods are designed by nature to turn a relatively small animal into a very big one in a very short amount of time. They will do the same to you.

- Limit your consumption of lean meats, such as fish, chicken, turkey, or range-fed beef, to a 3½-ounce serving. It is permissible to eat that size serving every day, but for more rapid weight loss, try to limit these foods to three or four times per week.
- Walk for at least 30 minutes a day, at least 6 days a week.
- Follow the additional guidelines that follow. (The menus and recipes in Chapters 9 and 10 will show you what to choose on a daily basis.)

A friend of mine is a master chef of natural-foods cuisine and cooks for an Academy Award–winning actress. The diet he prepares for this woman is essentially macrobiotic. One day recently he came to me and told me that his client had gained a little weight and wanted to lose it.

"What do you think I should do?" he asked me.

"Reduce the amount of grain you're giving her," I said.

"But grain is the center of the diet," he said. "It's the foundation of health."

"Increase her vegetables and reduce the grain. She'll never be hungry and she'll lose weight, fast," I said.

"You think?" he said, as if he was violating some sacred principle.

"Dude, if she doesn't lose weight, she'll find someone else who'll give her the results she wants," I said. "If you give her more vegetables, her health will get better, she'll get the results she wants, and she'll keep you as her chef."

My friend did exactly that and his client lost all the weight she wanted. She was thrilled and renewed his contract.

My point in telling this story is threefold:

- First, don't get locked into the dogma of any program. Instead, eat the foods that will accomplish two goals at once: rapid weight loss and health promotion.
- Second, recognize that the single greatest food for achieving both rapid weight loss and health promotion are vegetables.
- Third, reduce grain if you are already on a healthy diet and want to lose more weight.

The program described in Chapter 4 can cause you healthful and rapid weight loss. It can also eliminate many chronic and nagging symptoms that you may be experiencing now. But here are some additional tips that may speed weight loss and promote rapid improvements in your health.

Cooked vegetables are the secret to losing weight. The reason, very simply, is that they are exceedingly low in calories. A 3½-ounce serving of cooked broccoli (about 1 cup) provides only 28 calories. A pound of broccoli—which would be exceedingly difficult for you to eat—provides only 130 calories. A cup of cooked cabbage (again, 3½ ounces) provides only 24 calories. The same serving of carrots, 43 calories; cauliflower, 25 calories; kale, 50 calories; sweet corn, 86 calories; onions, 38 calories; dark green romaine lettuce, 16 calories.

Vegetables possess a unique combination of characteristics that make them perfect for losing weight.

- First, they are low in calories.
- Second, they are high in fiber and water, neither of which contains calories.
- Third, their low calorie and high fiber and water content allow you to eat them until you are full. You will still lose weight. In fact, you can eat vegetables throughout the day and continue to lose weight.
- Finally, vegetables are rich in nutrition, which means that while they promote weight loss, they also promote good health.

I recommend that you combine the green and leafy vegetables, which cause the greatest amount of weight loss, with the vegetables that provide the greatest feelings of fullness and satiety, which are the pulpy, sweet vegetables. Even the pulpiest and most filling vegetables—the potatoes, sweet potatoes, and squash—are still very low in calories. A big potato provides less than 200 calories. Three-and-a-half ounces of butternut squash provides only 45 calories. (It's worth noting that accompanying those 45 calories are also 7800 IUs of vitamin A, vitamin B_6, beta carotene, vitamin C, folic acid, magnesium, manganese, potassium, complex carbohydrates, protein, and fiber. There's less than 1 gram of fat in a serving of butternut squash.)

For maximum weight loss, I recommend at least two servings of green and leafy vegetables per day, and preferably three or four servings. And at least one serving of sweet vegetables, and preferably two. You can add fruit and the quick-cooking grains for greater satiety and satisfaction. Also, one serving of beans or lean animal foods can also be eaten. Beans are luscious, satisfying, and filling. They are also rich in phytochemicals, including phytoestrogens, which help to prevent cancer.

Vegetables That Speed Weight Loss

Here are examples of the green and leafy vegetables for maximum weight loss:

- Artichokes
- Asparagus
- Bamboo shoots
- Beets
- Beet greens
- Broccoli
- Brussels sprouts
- Burdock
- Cabbage
- Carrots
- Carrot tops
- Celery
- Chicory root
- Chinese (Napa) cabbage
- Collard greens
- Cucumbers
- Curly dock
- Daikon radishes
- Dandelion greens
- Dandelion root
- Endive
- Escarole
- Green peas
- Jinenjo potatoes
- Kale
- Kohlrabi
- Lamb's quarters
- Leek greens
- Leeks
- Lettuce
- Lotus root
- Mustard greens
- Okra
- Onions
- Parsnips
- Parsley
- Plantain
- Red radish
- Rutabagas
- Scallions
- Shepherd's purse
- Shiitake mushrooms
- Snow peas
- Sorrel
- Sprouts
- String beans
- Swiss chard
- Turnips
- Turnip greens

Vegetables That Provide Satiety and Satisfaction

The vegetables that will provide the greatest satiety, or feelings of fullness and satisfaction, are the pulpy, starchy vegetables, such as the following:

- Squash:
  - Acorn
  - Buttercup
  - Butternut
  - Hokkaido pumpkin
  - Hubbard squash
- Pumpkin
- Turnip
- Rutabaga
- Sweet potato
- Yam

Recommendations for Oil

For those who like to sauté vegetables in oil, but want to lose weight, I urge you to use oil sparingly. Remember, oil is liquid fat, which means it's rich in calories. A tablespoon of oil contains 125 calories. Limit your oil and combine it with water to make the oil go farther. (See Chapter 10.) Eat oil only two times per week.

Whole Grains for Weight Loss

In addition to vegetables, use the quick-cooking grains, which are easily prepared and low in calories; and occasionally beans, especially the low-calorie beans.

The grains for regular use are:

- Amaranth
- Buckwheat or kasha
- Millet

Oats, both rolled and steel cut
Quinoa
Teff

These should be your main grains. Eat them twice a day in small quantities. Supplement them with occasional use of brown rice and barley.

Once or twice a week, you can add pasta with green and leafy vegetables. Prepare and serve the pasta without oil. The vegetables will make the meal filling and satisfying without the calories, and without having to fill up on more pasta.

Beans for Weight Loss

In addition to the whole grains, eat beans four or five times per week.

The beans for weight loss and regular use are:

Aduki beans
Black beans
Chickpeas
Lentils

Animal Foods for Weight Loss

The best forms of animal foods for weight loss are fish, the white meat of chicken, and lean meats. Limit the amount of animal foods you eat to 3½-ounce servings (about the size of a deck of cards). Follow the guidelines for animal foods consumption in Chapter 5. You can eat a small amount of animal foods every day, but you will experience more rapid weight loss if you limit animal foods to three or four times per week.

Fruit for Weight Loss

One serving of any fruit per day will promote satiety and fullness. All fruits are fine, except avocados, which are higher in fat.

Weight loss is easy if you follow the recommendations I provide. The added benefit is that if you follow the diet I describe, you will feel and look more beautiful than you ever dreamed possible.

Overcome Allergies

One of the first things people experience when they adopt the Energy Balance program is elimination of allergies. The reason is because the program eliminates the primary cause of allergies, which is dairy foods. The proteins in cow's milk cause an immune reaction that can trigger any number of overt symptoms. For sensitive people, especially children, the most common reactions to milk, yogurt, and cheese are asthma, digestive disorders, chronic runny nose, bedwetting, ear infections, and eczema.

As I pointed out in Chapter 4, for certain sensitive children, dairy foods are especially dangerous because they can cause juvenile diabetes. Dairy proteins have also been shown to trigger an immune reaction that can damage the kidneys.

Researchers have found that dairy proteins can cause anemia is young children, as well. Cow's milk blocks iron absorption. Without adequate iron, the body cannot produce sufficient hemoglobin, which carries oxygen inside the blood. The result is anemia, which in turn produces fatigue. The proteins in cow's milk can also irritate the lining of the intestinal tract and cause microscopic bleeding, which results in further losses of iron and a greater likelihood of anemia.

Another food that I routinely ask people to eliminate, especially when they suffer from allergies, is wheat and other glutinous grains, such as corn and sweet rice. All of these grains can trigger allergic reactions. Many people with type O blood have trouble digesting wheat. Wheat contains lectins, a form of protein that acts as a binding agent, or glue, which allows substances to bind to one another inside the body. Lectins are used by organs, the immune system, bacteria, and viruses to bind to cells and tissues, such as the mucous membranes inside the intestinal tract. Food contains proteins that act as lectins, as well.

Food lectins can be incompatible, and even antagonistic, to specific blood types, says Dr. Peter J. D'Adamo, author of the book, *Eat Right for Your Type* (Putnam, 1997). When you eat a food that is incompatible with your system, the lectins can collect, or agglutinate, inside specific organs, such as the kidneys, intestinal tract, and liver. Once there, they can trigger immune and inflammatory reactions that can cause pain, swelling, and a reduction in the organ's function. The lectins in wheat and other glutinous grains can cause this very reaction in the intestinal tract of type O people, D'Adamo says.

I frequently take people with allergies off wheat and find that their immune systems calm down and their allergies go away. Other foods that can trigger an allergic reaction include eggs, chocolate, and citrus fruits.

In order to rid yourself of allergies, therefore, eat the healthy eight and eliminate all milk products (including milk, yogurt, and cheese), wheat, corn, and the more glutinous form of brown rice, known as sweet rice. Also avoid eggs, chocolate, and citrus.

Overcome Diabetes

A February 2002 issue of *The New England Journal of Medicine* reported a study in which researchers compared the effects of diet and exercise versus a drug treatment for people at high risk of getting diabetes. As you might expect, the diet and exercise reduced the risk of getting diabetes to a far greater extent than the drug did. No drug ever created can duplicate the effects of appropriate diet and exercise on health. This is especially true in the case of adult-onset (type II) diabetes.

Why? Because adult-onset diabetes is caused by red meat, dairy products, and processed foods. That's the conclusion of an enormous body of scientific research that started to accumulate at the beginning of the twentieth century. It was most recently confirmed in a study published in the February 2002 issue of the *Annals of Internal Medicine*. The research has consistently shown that a diet rich in fat and processed foods causes diabetes. The same *Annals of Internal of Medicine* study also found that a diet composed of whole grains, vegetables, fruit, fish, and poultry significantly reduce the chances of contracting diabetes.

Even the American Diabetics Association recommends a diet rich in grains, vegetables, beans, and fruit to treat diabetes.

Sixteen million Americans suffer from diabetes. Ninety percent of them have type II diabetes. The formula for getting type II diabetes is the standard American diet, plus a sedentary lifestyle, plus excess weight.

The New England Journal of Medicine study reported that even modest changes in lifestyle could reduce the risk of getting the illness. All you have to do is reduce your weight by 7 percent and do 2½ hours of brisk walking per week (less than half an hour a day for 3 days). That's not much. But what it says to me is that the human body can overcome diabetes when it's given even the slightest opportunity to do so.

It's worth mentioning that 78 percent of the people who took the drug treatment suffered significant side effects, including diarrhea, flatulence, nausea, and vomiting.

Here's a formula for getting rid of type II diabetes—without the negative side effects:

1. Eat the healthy eight every day.
2. Exercise by walking a half-hour a day, six days a week.
3. Avoid processed foods.
4. Stop all dairy products.
5. Stop smoking, if you smoke.

Protect Your Heart

Among the primary causes of heart disease are excess dietary fat consumption, specifically saturated fats found in animal foods, red meat, pork, eggs, and dairy foods. Saturated fat raises blood cholesterol levels, which contributes to the creation of atherosclerosis.

There are four types of fats:

- **Polyunsaturated fats,** found in plant foods and vegetable oils. These appear to lower blood cholesterol somewhat. When consumed in moderation, and especially as part of plant foods, these oils are associated with lower rates of heart disease and cancer.

  Most polyunsaturated vegetable oils, including corn, sunflower, and safflower oils, are known as omega-6 fats, which, when consumed in excess, depress immune function and increase the risk of cancer.

  A type of polyunsaturated fat known as omega-3 oils are found in cold-water fish, such as salmon, cod, flounder, haddock, mackerel, tuna, scrod, and swordfish. These fats lower cholesterol, raise the heart-protective cholesterol, HDL (high-density lipoprotein), and boost immune function.

- **Monounsaturated fats** are found in olive oil, seeds, and nuts. These have little or no effect on the bad cholesterol, known as LDL cholesterol (low-density lipoprotein). Olive oil may raise the good cholesterol, HDL (high-density lipoprotein). Olive oil also contains antioxidants and phytochemicals that, when consumed in moderation, boost the immune system and the cancer-fighting forces, as well. Olive oil is associated with low rates of heart disease and cancer.

- **Saturated fats,** found in animal foods, especially red meat, dairy foods, and eggs, have been shown to raise blood cholesterol levels and increase the risk of heart disease, cancer, and adult-onset diabetes. Palm and coconut oils are both rich in saturated fats, as well.
- **Trans-fatty acids.** These are fats that are created when polyunsaturated fats are infused with hydrogen atoms to make them thicker, so that they can be used in baking, in processed foods, and as margarine. These "hydrogenated" fats, or "hydrogenated vegetable oils" are among the most damaging fats to the human body. They are totally artificial and are associated with higher rates of cancer, heart disease, and diabetes.

All forms of fat, as well as excess calories and sugar, raise blood fats, or triglycerides. About 20 percent of triglycerides turn into blood cholesterol. That means that the more you eat, or the more processed foods you eat, the higher your cholesterol level goes.

Once inside your blood, cholesterol becomes oxidized, or starts to decay. Certain cells in your immune system, called macrophages, recognize these decaying cholesterol particles as a threat to health. The immune cells rush in and start gobbling up the tiny bits of cholesterol. Unfortunately, the cholesterol particles destroy the macrophages. The dead macrophages become imbedded into the walls of your arteries, where they form a yellowish, fatty streak. That's the first stage of atherosclerosis, or cholesterol plaque. If saturated fat continues to be consumed, the plaque gets bigger.

Plaques grow like boils inside the arteries. They contain liquid pools of cholesterol inside them. This makes them highly volatile and prone to bursting open, like a big pimple bursting open on your skin. One a plaque erupts, its contents come spewing out and a wound is formed inside your artery. Blood platelets and other clotting factors arrive on the scene and create a scab over the wound, much as they do on your skin. As the boil grows and erupts again and again, the scab gets bigger and bigger. Eventually, that scab can become a very large thrombosis, or clot, that can block blood from flowing inside your artery.

If an artery that brings blood to your heart gets blocked, a part of the heart suffocates and dies, an event known as a heart attack. If that same thing happens in an artery that leads to your brain, a part of your brain dies and you suffer a stroke.

The good news is that these boils can be made to shrink and it can happen fast. If you drastically reduce or eliminate saturated fat from your diet, and eliminate processed foods, your blood cholesterol level will fall very quickly. When that happens, the cholesterol inside these plaques starts to drain. This stabilizes the boils and prevents them from erupting. Gradually, the boils start to shrink. Eventually, they get so small that they are no longer a threat to your health.

Reversal of atherosclerosis, or the shrinking of the cholesterol plaques, was demonstrated in a study by Dean Ornish, M.D., and his colleagues and published in the July 21, 1990, issue of *The Lancet* (336:129, 1990). Ornish showed that as the plaques got smaller, the symptoms related to heart disease, such as chest pain, also diminished. As I said, these benefits can happen quickly. Ornish showed that people experience a reduction in the size of the plaque and a significant reduction in chest pain within one month on a proper diet.

Here's a formula for preventing heart disease and curing it:

1. Eat the healthy eight every day.
2. Eliminate processed foods.
3. Eliminate dairy foods.
4. Limit animal food consumption to 3½-ounce servings of fish (your first choice) or lean meat, such as chicken (your second choice), and eat those foods only 4 days per week.
5. Walk 6 days a week for at least 30 minutes a day.

This will bring your cholesterol down below 180 mg/dl and help you reverse cardiovascular disease.

Prevent Breast and Prostate Cancer

Researchers now realize that breast and prostate cancer are essentially the same disease, except that breast cancer happens more frequently in women, and prostate cancer exclusively in men. Both diseases are caused when reproductive hormones, specifically estrogen in women and a form of testosterone in men, become elevated and remain excessively high over time.

The primary culprit in both cases is animal foods, especially fat, which drives up reproductive hormones and keeps them elevated. Interestingly, about a third of men have prostate cancer by the age of 60, and by 80, as much as 80 percent have it. However, a great percentage of these men never know they have prostate cancer. They live active lives, experience no symptoms, and die in old age. Researchers have found that the central difference between those men who get prostate cancer that does not spread, or cause symptoms, versus those who get prostate cancer and have it metastasize and eventually kill them is diet, especially a diet rich in fat and animal protein.

In Scandinavia, where the diets are rich in meat and dairy foods, prostate cancer is far more prevalent than in Asia, where the diets are low in animal foods and rich in plant foods. Men in Sweden are eight times more likely to suffer from prostate cancer than men in Singapore, for example.

Autopsies on men in Japan, for example, consistently reveal latent prostate cancer that did not spread or manifest symptoms. When Japanese men come to the United States and adopt more Western-like diets, their rates of malignant prostate cancers go up to American levels.

Similar kinds of patterns exist for breast cancer. Women in Asia experience far less breast cancer than women in the West. And when women in Asia get breast cancer, they have a much higher rate of survival than women in the United States. But when Japanese women migrate to the United States and adopt diets similar to those of Americans, their rates of breast cancer rise to American levels.

There are several factors that combine to create high levels of breast and prostate cancer. They are the following:

1. High consumption of fats from animal foods
2. Overweight
3. Low intake of plant-based chemicals, such as phytoestrogens
4. Low fiber content of the diet
5. Lack of exercise
6. High consumption of alcohol

Let's look at these six factors individually, especially in relationship to breast cancer.

High Consumption of Fat from Animal Foods

Aging and degenerative diseases, such as cancer, have the same cause, namely oxidants, also known as free radicals. Oxidants are highly reactive oxygen molecules that cause the breakdown of molecules that make up our cells. As these molecules decay, they cause cells and tissues to become deformed. Many of our tissues and organs start to shrink, shrivel up, and scar. Oxidants cause our skin to wrinkle and shrivel up as we age. They do this by drying out and shrinking the collagen foundation of our skin. Once that happens, the surface skin is forced to fold over on itself and wrinkle.

Sometimes oxidants interact with our DNA, killing some cells and causing others to become cancerous. Oxidants are the primary cause of such diseases as heart disease, cancer, arthritis, Alzheimer's, Parkinson's, cataracts, glaucoma, immune diseases, and kidney and liver disorders, such as cirrhosis.

The primary causes of oxidants are cigarette smoking, dietary fat, alcohol, processed foods, x-rays, the sun's ultraviolet rays, industrial pollutants, and chemical additives in our foods. To a great extent, we determine the level of oxidation in our tissues by our diets (a high-fat diet is one of the principle sources of oxidants), our smoking habits, the level of pollution in our environments, and our exposure to ultraviolet radiation from the sun. That means that we determine to a great extent the speed with which we age and our chances of contracting a major illness.

The antidote to oxidants are antioxidants, which slow or stop oxidation and the breakdown of molecules, cells, tissues, and organs. The primary antioxidants are vitamin C, E, and beta carotene, all of which are found in plant foods. Other antioxidants include vitamin B_6, the mineral selenium, magnesium, and glutathione, found in most plant foods. The most abundant source of antioxidants are plant foods. There are really no significant quantities of antioxidants found in animal products.

Just as our diets can determine the levels of oxidants in our systems, so do our diets determine our antioxidant levels.

Oxidants are what cause cancer, including cancers of the breast, colon, and prostate. Excess amounts of amounts of dietary fat, especially from animal foods, triggers the production of oxidants, which in turn can deform DNA and set malignant cells in motion.

Not only do oxidants trigger the origin of cancer, they also nourish cancer once it manifests. Therefore, fat plays a role in both the creation and the nourishing of cancer cells and tumors. And it doesn't stop there.

Overweight Promotes Cancer

When it comes to cancer, one of the big concerns to both men and women is overweight. Fat cells are estrogen factories. They also elevate a type of testosterone called dihydrotestosterone. When these hormones become elevated, they overstimulate the hormone-sensitive organs. In women, those organs are the breast, ovaries, and uterus. In men, the testes and prostate. With continued overstimulation, cells and tissues can be deformed. In some cases, they can become cancerous.

High estrogen levels in women cause fibrocystic breast disease, heavy menstrual bleeding, all the symptoms related to premenstrual syndrome (PMS), uterine fibroids, and cancers of the breast and uterus. In men, high levels of dihydrotestosterone can give rise to prostate enlargement and prostate cancer. But the origin of all these conditions is the rich Western diet, and especially dietary fat.

Not only does overweight trigger the underlying causes of cancer, but it fuels cancer once it arises. This means that overweight people have a much greater chance of dying once a cancer manifests. A study known as the Iowa Women's Health Study follows the health patterns and diets of 698 postmenopausal women with breast cancer. The researchers found that overweight women are nearly twice as likely to die of breast cancer than lean women. They also discovered that women with the highest dietary fat intake also have twice the risk of dying of breast cancer.

Excess weight puts a woman's life in jeopardy, especially from breast cancer. Women in their 50s who gained 11 pounds or more during the previous 10 years had twice the risk of developing breast cancer than women who remained lean, according to a study done by Regina Ziegler, Ph.D., MPH, of the National Cancer Institute, and reported in the May 15, 1996, *Journal of the National Cancer Institute*.

If you are overweight, follow the Energy Balance Diet to lose weight and dramatically reduce your risk of cancer.

Low Intake of Plant-Based Chemicals, Such as Phytoestrogens

Women who eat five or more servings of vegetables a day have half the risk of developing breast cancer than those who ate fewer than three servings each day, according to a study published in the March 20, 1996 issue of the *Journal of the National Cancer Institute.*

Plant foods protect us from just about every health problem out there today, especially from cancer. The cruciferous vegetables, specifically broccoli, cabbage, kale, Brussels sprouts, collard greens, mustard greens, and watercress, contain compounds known as indoles. These chemicals inhibit estrogens that cause tumors to form in the breast, thus protecting breast tissue from cysts and cancer, researchers have found.

Another substance found in these vegetables—and that is especially abundant in broccoli—is a chemical called sulforanphane, which detoxifies the blood and tissues. Sulforanphane triggers the production of cancer-preventing enzymes in the blood, according to a study published in the *Procedures of the National Academy of Sciences* (vol. 89, March 1992).

Phytoestrogens, or plant estrogens, protect women from breast cancer and men from prostate cancer. Each cell has estrogen-receptor sites located on its cell membrane, much like doors on a room. When we eat a diet rich in plant estrogens, these chemicals attach themselves to the doors, thus preventing the cancer-causing estrogens to enter the cells.

Another phytoestrogen that is especially powerful at protecting against cancer is a substance known as genistein, which blocks blood vessels from attaching to tumors, scientists have discovered. Tumors need oxygen and nutrition to survive, just like healthy cells do. Genistein blocks blood vessels from attaching themselves to tumors, and thus prevents the tumors from getting what they need to survive. This discovery, which was reported in the April 1993 issue of *Proceedings of the National Academy of Sciences*, is now being explored by scientists as a way of preventing and treating various types of cancers, including those of the breast, prostate, and brain.

Genistein is abundant in soybean products, including the beans, tempeh, tofu, shoyu, tamari, miso, and natto (a soybean condiment).

Low Fiber Intake

Fiber binds with estrogen and eliminates it from the body through the feces. So the more fiber in your diet, the lower your estrogen levels. A study published in *The New England Journal of Medicine* showed that women on vegetarian diets eliminate two to three times more estrogen in their feces than nonvegetarians. Another study showed that a diet rich in fiber and low in fat reduced estrogen levels in postmenopausal women by 50 percent!

Researchers have found that a 17 percent drop in estrogen levels reduced the risk of breast cancer fourfold to fivefold. That means that if you're a woman and want to protect yourself from breast cancer, you've got to start by eating a high-fiber diet. Vegetables, grains, beans, fruit, nuts, and seeds are the only sources of fiber in the food supply. Animal foods have no fiber.

Not only does fiber protect a woman from breast cancer, but by lowering estrogen, fiber reduces PMS symptoms. That's why I told the woman I wrote about earlier to increase her intake of plant foods, because the fiber alone would reduce her PMS symptoms.

Lack of Exercise

Studies have consistently shown that women who exercise regularly—at least four days a week—have significantly lower rates of breast cancer than women who do not exercise.

Alcohol Intake

Alcohol raises estrogen levels in both women and men, thus increasing the risk of breast and prostate cancer. A woman's chance of getting breast cancer increases 25 percent when she consumes two alcoholic drinks per day, studies have shown. If you drink, limit alcohol to three or four drinks per week.

Prevent Cancer

Here is a simple formula for protecting yourself against cancer:

- Eat the healthy eight every day.
- Walk and do some other form of aerobic exercise six days a week.

- Limit alcohol consumption to four or fewer drinks a week.
- Don't smoke.
- Avoid processed foods.
- Avoid artificial ingredients, such as pesticides, herbicides, and colors.
- Lose weight. (The other recommendations can achieve this goal without your having to make a conscious effort.)

In addition, use the hot towel scrub and other body care methods I describe in the following sections.

I have focused on breast and prostate cancers, the two most common cancers in men and women, but all the common cancers—including colon and lung cancers—arise from essentially the same causes, namely the standard American diet. If you follow the advice I have given for breast and prostate cancers, you'll also dramatically reduce your chances of getting colon, lung, pancreatic, and skin cancer.

Rub Your Body, Feet, and Tongue Clean

For many years, I have recommended that people use various kinds of body care methods to promote healthy circulation, elimination, and abundant energy. Here are just a few of the procedures that I urge you to try.

Hot Towel Scrub and Other Body Care Tools

I advise my clients to use a warm, wet towel to vigorously scrub the body, either while in the shower, or just after you get out of the shower and are still wet.

I call this the hot towel scrub. It promotes healthy circulation and the elimination of toxins from the skin. The primary blood-cleansing organs are the liver and kidneys. When toxins exceed the capacity of these organs to cleanse the blood, they attempt to eliminate them through the skin. Unfortunately, the more the skin is used to discharge waste products, the more likely the pores are to become blocked. The best way to unblock the pores, promote elimination through the skin, and relieve the stress on the kidneys and liver is to rub the skin clean. Scrubbing with a hot, wet towel will open the pores and rub away so much residual waste that has been building up on the skin's surface.

Scrub your entire body down with a warm, wet towel for ten minutes, at least three times a week. Not only will it promote healthy elimination of poisons, but it will make your skin feel smooth, soft, and youthful again.

The Tongue Rake

Two other helpful tools are the tongue rake (or tongue scraper), which can be used to scrape the tongue clean of its coating, and the foot file, which can be used to eliminate unwanted calluses. The tongue rake can wipe away the coating that prevents you from tasting your food. This, of course, will allow you to eat far more healthfully, because your taste buds will be free of so much build-up and consequently will be better able to taste the natural flavors in plant foods. Scraping your tongue will also help eliminate waste from the liver and intestinal tract, which is continually sending waste products up and out the mouth, especially in people who are chronically constipated.

File Away Calluses for Better Health

The foot file will eliminate calluses and promote better circulation and improved breathing in your feet. Like the wet towel scrub, the foot file will help eliminate waste products from the body. The foot file will also maintain the health of your feet.

But there are other reasons why I promote the use of the hot towel scrub, foot file, and other forms of bodywork, such as massage. Those reasons relate to more ancient and traditional forms of healing, especially Chinese medicine and Ayurveda (about which there is much more in the next chapter).

Caring for your body is an act of self-love and self-healing. You cannot expect your body to take care of you if you don't take care of it.

One Disease, Many Forms

If a single set of dietary recommendations address all of these illnesses, what does that tell us about these diseases? In fact, they're all the same disease. They only appear to be different because the underlying illness manifests in one person as heart disease, in another as cancer, and in still another as diabetes. But the underlying illness is the same in all three cases. And that illness isn't really an illness at all. It's poisoning. We are poisoning ourselves and calling it a disease.

Why doesn't the poisoning cause the same illness in everyone, you might ask. That would be a good question. The answer is that we all have our own unique genetic strengths and weaknesses. As the toxic levels cross a certain threshold, they trigger the genetic illness that the person is most susceptible to. Some people have a genetic weakness for heart disease, others for colon cancer, some for breast cancer, others for Alzheimer's disease. The list goes on and on.

There are vast differences in treatments for each of these illnesses, but they really have the same causes and ultimately the same cure. As long as we don't recognize that we are poisoning ourselves, we continue to think that these illnesses mysteriously arose from outside of us, or from a genetic flaw in our makeup. But if you realized that you were poisoning yourself between four and nine times a day, you would marvel at your body's capacity to withstand the terrible toxic load it is exposed to each day.

Follow the instructions in this chapter and watch how your body responds. In a very short time, you very well could experience a dramatic improvement in your health. With that improvement could come the realization that you are much more in command of your own life than you ever dreamed possible.

chapter 7

Ancient Medicine for Maximum Beauty

One of the underappreciated tragedies of our culture is that we have come to see only one body type as truly beautiful, especially when it comes to women. That type is tall, lean, and thin-boned. That rather arbitrary perspective makes it extremely difficult for a lot of women and men to be seen as beautiful. Did you know that in 2002, the average dress size in America is a size 14? (In 1985, it was a size 8.) The average model size is a 4. That means that a lot of women today do not fit the cultural definition of beauty. To be sure, an equal number of men don't fit the prevailing definition of beauty, either. What's worse, however, is that so many women and men who have larger bodies are attempting to diet themselves into smaller sizes so that they can fit today's definition of beautiful. Most, as we know, are failing, despite continual efforts.

Two things are working against them. First, the standard Western diet makes it almost impossible to lose weight, unless you are willing to starve yourself. (Many are, which is why we suffer from such high rates of eating disorders.) Second, they have constitutions that are very different from the thin, small-boned bodies that are the current cultural definition of beauty. People with thicker, stronger bones, and wider body shapes are never going to look like today's supermodels. That does not mean, however, that they are not beautiful, or that they are incapable of becoming beautiful versions of themselves. What it does mean is that too many people are trying to become something they are not—and never will be, no matter how hard they try.

At other times in history, the definitions of beauty were more varied than our own. All you have to do is look at the paintings of the seventeenth-century Flemish artist Peter Paul Rubens to know that larger women were once regarded as the standard form for beauty.

The real key to beauty is—and always has been—health. People who are healthy have all the attributes of good beauty, meaning bright and healthy skin, abundant and radiant energy, optimism, the ability to enjoy life, and a capacity to give and receive love. As everyone knows, people of all different body types can have these qualities. And we have all known people of every shape and size whom we recognized as beautiful. Therefore, the real question facing us is how to achieve better health, from which beauty will naturally flow.

During the past two decades, people have been attempting to find new approaches to achieving better health and more beauty through complementary forms of medicine. Among the healing modalities that people are turning to the most are traditional Chinese medicine and Ayurvedic healing. Both of these approaches give us new ways to see the body and achieve better health and optimal weight. In this chapter, I want to introduce you to two sets of ideas that may help you achieve a better understanding of your body type, as well as give you another set of tools to achieve better health and optimal weight. The first of these is the Ayurvedic understanding of body type. The second is the Chinese approach to weight loss. Let's look at these one at a time.

Ayurveda: Three Beautiful Body Types

Three thousand years ago, the sages of India created the healing system known as Ayurveda. Ayurveda teaches that there are at least three different body types, referred to as Vata, Pitta, and Kapha; each is distinguished by its unique size, shape, and psychological nature. All three have their strengths and weaknesses; each is associated with certain personality characteristics; and when it comes to healing, each is treated a little differently. Needless to say, all three have the potential to be beautiful. The critical issue for all three is not to be like each other, but to be healthful versions of themselves.

Both Ayurveda and traditional Chinese medicine (TCM) saw beauty as a by-product of health. What we in the West refer to as light in the eyes and radiant skin, traditional healers saw as the outward signs of a strong life force. All health was seen as dependent upon that life force. The stronger the life force, the more likely a person would experience health and beauty. Every traditional healing art used various methods, including diet, exercise, massage, acupuncture, and meditation, to promote and strengthen the life force. When the life force is strong, it can overcome any illness, no matter how serious.

In ancient India, that life force is known as prana; in China, it is known as chi; in Japan, ki; in ancient Greece, pneuma. The life force animates the body and makes it possible for every cell, tissue, and organ to function optimally. A person who radiates with energy and aliveness is both healthy and beautiful.

A person who is truly alive and healthy—by that I mean, overflowing with life force—is beautiful. You can see it in her skin, smile, eyes, and the way she moves. Her beauty announces itself. It is a force, a kind of power, whose origin is energy itself. People who are beautiful can have big bodies or small bodies, just as people who are ugly can have big bodies or small bodies. Size doesn't matter when it comes to beauty. Beauty has less to do with weight than it does with the power of life energy that radiates from every cell of the body and directs the body in its every movement.

The advice from a traditional healer to a client who wanted to be beautiful was simple: Become a healthy version of yourself, and you will be beautiful. Implicit in that advice is the requirement that you be yourself, honor your own body type, and never, under any circumstances, attempt to become something you are not.

In order to help you better understand your own body type and learn some additional ways to promote your own health and beauty, I have described the three Ayurvedic body types.

Essentially, the vata body type is what we imagine when we think of the current definition of beauty: thin-boned, tall and thin, or short and petite, and more cerebral in nature. The pitta type is more athletic, more muscular, and more fiery. The kapha type tends to be physically stronger, big-boned, full-bodied, with a more balanced temperament. It was the Kapha women that Rubens immortalized as the seventeenth-century image of beauty. Here's a more in-depth understanding of all three, as well as the kinds of foods and behaviors that promote the health of each.

Vata: Lightly Touching the Earth

Vata types are typically underweight and have difficulty gaining weight. They can have bony bodies—the actress Calista Flockart, who plays Ally McBeal on the television show of the same name, is a good example—and, for many, their joints protrude. Their metabolisms function very rapidly, burning up calories and even consuming any excess that the body may be trying to hold. Consequently, their bodies have little padding, so to speak, and often appear exposed and vulnerable, no matter what they may be wearing.

Vatas tend to be very intelligent, philosophical, and even spiritual in their orientation. Because of their inherently sensitive natures, many poets and writers are vatas. Their sharp minds make it easy for them to learn even difficult subjects quickly. They tend to get enthusiastic whenever they have the opportunity to learn something new.

Vatas' intellectual nature often makes them dreamy and caught up with their thoughts. Their minds are so active, in fact, that they tend to draw a great deal of the body's energy upward, into the head, where it is burned as fuel for their ever-active minds. In the competition between the mind and the body, the minds of vatas always win. For this reason, vatas are often said to live in their heads, rather than in their instincts and in the physical world. Because their minds are so active, their nervous systems can become overstressed and frazzled. They can be fearful, anxious, indecisive, and prone to overexcitement. Woody Allen is a vata type.

Vatas' nervous energy can make it difficult for vatas to sleep deeply. They frequently suffer from insomnia. When they do sleep, vatas tend to need fewer hours than others, often as little as four hours, or as much as six.

Most vatas do not enjoy athletics or strenuous physical activities. Their bodies are often uncoordinated and a little gawky. On the other hand, exercise releases much of the nervous tension they feel and balances their conditions. Therefore, vatas need regular exercise. The exercise that is most compatible with vatas is yoga, which is meditative (vatas are very spiritually oriented), gentle (vatas dislike highly active and violent games), and calming.

Vatas often have irregular features and prominent noses. Their hair is often dry and wiry, their skin often pale, cool, and sometimes rough. Vatas commonly have small, narrow eyes that are sensitive to light. Their eyes are often irritated and itchy.

Their weakest organs are their intestines. They suffer from poor nutrient assimilation—one of the reasons they can eat so much and remain thin—and are prone to constipation. Because they have small, thin bones, vata women must protect themselves from osteoporosis. Therefore, they must be careful about their protein intake. Animal foods make vatas feel grounded and centered—and therefore are essential for them—but too much can work against them by weakening their bones. A high-plant-food diet will keep this from happening. (See the following sections.)

Many vatas consciously or unconsciously know that they don't have a lot of excess energy, or reserves, to draw upon. This awareness makes them very cautious and self-protective, especially in intimate relationships. They don't want people robbing them of the energy they need to function and feel safe in the world. On the other hand, vatas are excellent counselors. They have insight into people and life and are able to communicate their ideas very well.

Vatas love the sun and the beauty of nature. They feel most relaxed and at home in their bodies when they are communing with nature.

Foods That Promote the Health and Beauty of the Vata Type

In general, vatas need foods that provide excess to their bodies, or what we might term nutritive foods. Their diets must have an abundance of cooked vegetables, which promote healthy assimilation and bowel function, but vatas benefit by eating regular amounts of animal foods, as well. Animal foods give them a sense of their own personal power. They make them more grounded in their bodies, more present in the moment, more confident and secure. Fish, low-fat meats, and eggs in small quantities are all ideal. Vatas can have a small quantity of animal food (3½ ounces), four or five times per week (that includes the eggs). This will give them energy, higher dopamine, and more feelings of personal power and physical stability.

Vatas must be careful about overdoing animal foods. Animal foods are difficult to digest and will stress the vata's already-weak intestines, especially the colon. With too much animal food will come more constipation. Also, the contracting influence of animal foods will drive the vata person toward sugar and other expansive foods, which will make them even more intellectual, more dreamy, and more out of their bodies. As with everyone else, the vata type must find his own balance between protein foods and carbohydrates, or animal foods and plant foods.

Vatas feel better when eating cooked food. Cooking makes the nutrients more accessible to their intestinal tract. Cooked food also warms the blood, and their generally cool condition.

The grains that are especially good for vatas are brown rice, oatmeal, and wheat. The vegetables: cooked asparagus, beets, carrots, sweet potatoes, radishes, and zucchini. The fruit: sweet fruits, bananas, coconuts, apples, figs, grapefruits, grapes, mangoes, melons (sweet), oranges, papayas, peaches, pineapples, plums, berries, and cherries. All the beans are generally good for vatas (very nutritive), but mung beans and black beans are especially good. Dried fruit is very supportive to the vata constitution, as are spices and sesame and olive oil.

All of these foods promote feelings of well-being, stability, and optimal function within the vata body type. The more grounded, healthy, and positive a vata type feels, the more beautiful he or she becomes.

Vata Body Type Strengths and Weaknesses

Strengths:

Can eat just about anything and not gain weight.

Highly intelligent, visionary, philosophical.

Thoughtful, sensitive, poetic nature.

Excellent counselors, advisors.

Weaknesses:

The primary imbalance is to bring energy up to the head, and to diminish energy in the body. Must get regular exercise in order to feel strong, secure, and present in the body.

Thin bones; must guard against osteoporosis.

Intestines tend to be weak; must protect against constipation and other bowel disorders.

Cannot tolerate much stress; they easily become anxious, nervous, and fearful.

Pitta: The Fiery One

The pitta body type is muscular and athletic, which makes pittas more physically oriented. They have yellowish or reddish skin that is sensitive to rashes and often freckled. They often sweat profusely. Their hair is susceptible to premature graying or baldness. Their bodies tend to be hot, which makes them prefer cooler temperatures. Heat irritates them. They often like hot, spicy foods, and alcohol, which is consistent with their fiery temperament. When they gain weight, it is usually evenly distributed throughout the body.

Pittas tend to be leaders. They are often well organized, highly intelligent, and charismatic. They can affect a superior and even arrogant air. They are usually emotional, competitive, passionate, and even hot-tempered. They tend to sleep on their backs and usually need eight hours' sleep to feel fully rested.

They have enormous appetites for food and for experience. When dealing with intense and chronic stress, they can become gluttonous and promiscuous. Their weaknesses are their liver, heart, and stomach. They can suffer from chronic diarrhea and insomnia.

Pittas have a constitutional talent for creating balance, both in their diets and their health. Although we all know intuitively what to eat, pittas are often more aware of what they need to eat to support their health. In the same way, they also are able to naturally engage in behaviors that balance their bodies—resting when they are overworked, exercising when they need more movement and strenuous activity. Because this intuitive understanding for balance comes naturally to pittas, they are often bored with balance. It comes too easy and it doesn't give them the excitement and instability they are looking for in order to feel that they are truly living. Consequently, pittas often seek out conflict in relationships and look for the issues between themselves and others to work on. That gives them the excitement they need.

Pittas need physical exercise. Their athletic bodies rapidly degenerate when they are sedentary. They need a physical outlet for their emotional natures. The combination of high emotion and sedentary lifestyle is especially destructive to the pitta constitution.

Foods for Pitta Constitution

Pittas need to chill. They need to let their natural gift for balance rule their lives to a greater extent, especially when it comes to their diets. In general, they should reduce extreme foods, especially hot spices and animal foods, and choose more cooling, balancing foods. The spices should be mild, such as coriander, cinnamon, cardamom, fennel, turmeric, and little black pepper. The grains should be barley, oats, rice (including basmati and white rice), and wheat. The vegetables should include sweet and bitter vegetables, including squash, pumpkin, asparagus, cabbage, cucumber, cauliflower, celery, French beans, lettuce, peas, parsley, potatoes, zucchini, sprouts, and watercress. The fruit: apples, avocadoes, coconuts, figs, melons (sweet), oranges (sweet), pears, plums, pomegranates, and dried plums.

These foods promote the life force of the pitta body type, and thus strengthen and enhance its health and beauty.

Because they are so athletic, the pitta body type needs regular exercise. If they don't get exercise, their emotional nature becomes unsettled and troublesome. They begin to see problems where they really don't exist. In addition, meditation, bodywork, and yoga are extremely helpful to the pitta body in order sustain its health and maintain a balanced approach to life.

Pitta Body Type Strengths and Weaknesses

Strengths:

Strong constitution.

Well-developed muscular system.

Athletic.

Natural talent for creating balance, both in themselves and in others.

Natural healers.

Leader types.

Weaknesses:

Their balanced natures make them seek out and create imbalance, usually in relationships.

Must eat and live so that they protect their heart, liver, and stomach. Can easily develop illnesses in these areas.

Very often tense, hyperactive. Have difficulty relaxing.

Highly emotional, often suffering outbursts of emotion.

Kapha: Standing Firm and Connected to the Earth

Kapha body types are solid, thick-boned, and strong. They have wide bodies, hips, and buttocks. Their thighs tend to be big and strong. Because they have such strong skeletons, they do not have to worry too much about osteoporosis. Kapha types are usually slightly overweight, or even considerably overweight.

The skin of kapha people is often pale and cool. Their hair is often wavy and abundant. The eyes are large, clear, and usually dark. Their natures are calm, peaceful, and methodical. They move slowly and deliberately. Kapha types have a stronger center of gravity.

When a kapha type sits down, he or she really lands. You feel as if they are truly present, truly taking up the space. Unlike vatas, who can be waify and easily overlooked, you cannot overlook a kapha type, even if he or she is not particularly overweight or tall. It's as if some huge invisible mass fully lands when a kapha comes to rest. They are planted and you know it. You realize that it would be difficult to move them physically, even if they are not particularly big or overweight. Kapha types radiate competence, even when they are quiet or shy.

Their excellent native intestinal strength makes it possible for them to easily assimilate nutrients, which is why they do not have to eat much to get an abundance of nutrition and calories, and why they are usually overweight. Kapha types also have slow metabolisms. This makes them slower of movement and more relaxed as people. However, these characteristics make it very easy for kapha types to become obese if they do not take care of their diets and exercise patterns.

Kapha people are practical and competent. They have a natural sense of how things work. They love the earth and their material possessions. They are sociable and enjoy their family and friends. They would prefer to be partiers, but they're usually hardworking.

Kapha people sleep deeply—they often snore—and need more than eight hours to feel rested. They can be stubborn and greedy.

Kapha types have so much life energy in their bodies, and such natural power, that they cannot help but be beautiful when they take care of themselves and eat properly. The foods that follow will enhance the health and beauty of the kapha body type.

Diet for Kaphas

Because kaphas have such strong digestion, they should not overeat. It will only make them heavier and more overweight. Kaphas should eat lots of vegetables and light foods. The richer the diet, the more likely they will suffer from heart disease, obesity, or cancer. Their heart is their most vulnerable organ.

Kapha types should eat barley, millet, corn, oats, and rice (including brown and basmati). The vegetables should include beets, cabbage, cauliflower, celery, eggplant, garlic, lettuce, mushrooms, onions, parsley, peas, radishes, spinach, and sprouts. Vegetables lighten and promote weight loss. Aduki and black beans are good for kaphas. The fruit should include: apples, berries, cherries, dried figs, mangoes, peaches, pears, raisins, and dried plums. The primary animal food should be eggs. All spices are good for kaphas. They should limit oil as much as possible.

All these foods will promote the health, vitality, life force, and beauty of the kapha body type.

In addition, the kapha body type needs to walk or do some other nonstrenuous but regular exercise. Kaphas are typically slow-moving and don't like such activities as running or highly vigorous sports. Yet they love to stroll, to enjoy nature, and to savor the moments. They need to walk daily in order to keep their weight down and sustain their fitness and health.

Kapha Body Type Strengths and Weaknesses

Strengths:

Strong constitutions.

Strong bones. Kaphas do not usually have to worry about osteoporosis.

Strong intestines and digestion. Great assimilation of nutrients.

Weaknesses:

Their great intestinal strength makes it easy for them to gain weight. They have to guard against overeating.

They can easily become obese if they do not take care of their diets and exercise patterns.

They can become couch potatoes and live sedentary lifestyles. They must guard against this pattern by getting plenty of exercise and watching their diets.

As far as Ayurveda is concerned, there is no body type that is superior to another. Rather, all three have their own unique strengths and weaknesses. If one body type is prized above the rest, it would be the kapha type, because they are regarded as having the strongest bones and intestinal tract, which makes them most capable of enduring the demands of life and producing offspring. Yet vata and pitta have their unique strengths, as well. From the traditional perspective, all three have the capacity for health and beauty, which is the foundation for a happy life.

Using Chinese Medicine to Promote Optimal Weight and Health

In Chapter 2, I described how the energy wave affects insulin, energy, and weight. From a Western medical perspective, the organ that is most effected by a fluctuating energy wave is the pancreas. In Chinese medicine, the pancreas and spleen are seen as related organs, but from the Chinese perspective the spleen has the superior role. The spleen, say the Chinese, is responsible for maintaining your body structure and your weight. Therefore, all behaviors that support the spleen will help you reduce weight and create a body shape that is most beautiful for you.

Nothing illustrates the unbreachable gulf between Western and Chinese medicine better than their respective understandings of the spleen. From a Western medical approach, the spleen is part of the lymph system and is seen as the medium through which immune cells enter the bloodstream. The spleen also filters broken red blood cells from the blood.

From the Chinese medical perspective, the spleen holds a very special place in the function of the human body. To begin with, the spleen distributes life force, or chi, to the entire digestive tract, as well as to the heart. The spleen is often referred to as the governor of digestion. The digestive organs, as well as the heart and to a lesser extent, the kidneys, depend on the spleen for their most basic form of nurturance, namely chi, or the living energy that animates and supports the health of the human body.

The spleen also has a central role in weight management. The Chinese say that the spleen maintains the integrity and shape of the overall body. It does this in three ways:

- First, by being responsible for the health and vitality of the small intestine, the organ that absorbs nutrients from our food
- Second, by sustaining the function of the large intestine, which is responsible for eliminating waste
- Third, by regulating metabolism, or the rate at which we burn calories

In these three ways, the spleen has a great deal to do with the size and shape of the body, and its overall integrity. Because of these and other responsibilities, many Chinese healers believe that the spleen is one of the most important organs in the body.

When one is chronically underweight, or chronically overweight, a Chinese healer strengthens and nourishes the spleen. When the spleen is brought back into balance and health, it will restore the natural size and shape of the overall body.

Typically, the spleen is either underactive, also referred to as deficient, or is overactive, sometimes referred to as excessive. Deficient spleen means that the organ is lethargic, or functioning more slowly and sluggishly. This will slow metabolism, which means weight is more likely to increase. Also, fat and waste products are likely to accumulate. A person who is chronically overweight may have a deficient spleen, meaning that the spleen is functioning slowly, sluggishly, and sub-optimally. Digestion is also moving more slowly.

An excessive, or overactive, spleen means that the organ is overactive, or racing. That means that metabolism is probably also racing. This will speed the burning of fuel and the utilization of nutrients. An overactive spleen can make someone chronically underweight.

The vata body type is the most susceptible to chronic underweight. Not surprisingly, the vata type frequently suffers from overactive spleen. The body type that is most likely to be overweight is the kapha type. When a kapha is overweight, his or her spleen is probably underactive.

How the Spleen Is Weakened

Chinese healers say that the spleen is a very delicate organ. It is especially sensitive to intense sweetness, sugar, acidic foods, and spices, which all injure the spleen. Lack of chewing injures the spleen, as well. Chewing increases the production of saliva, which is alkalizing. When the saliva is not adequately mixed into the food, the food becomes more acidic and injurious to the spleen.

Initially, these foods and conditions cause the spleen to become overactive. Overactive spleen is associated with frequent bowel movements, diarrhea, and indigestion. However, with time, the overstimulation of the spleen makes the organ sluggish and tired. Eventually it becomes bloated and weak, therefore preventing it from sending adequate life force, or energy, to the system at large, and to the digestive tract in particular. Among the consequences of a weak spleen are weak pancreatic function, overweight, weak digestion, and poor elimination. Many people are constipated, the Chinese point out, not because their diets lack fiber, but because they have eaten too much sugar, acidic foods, and strong spices, all of which have injured the spleen function and consequently weakened the large intestine.

All digestive issues are seen in Chinese medicine as related to the health of the spleen. All forms of gas—belching, flatulence, rumbling stomach—are spleen-related. So, too, are heartburn, acid indigestion, and nausea. These symptoms indicate a spleen imbalance—usually, an excess of energy that the spleen cannot distribute in an orderly fashion.

As the spleen weakens, the person's weight increases and the pancreas gets weaker and unable to produce adequate insulin. This sets the stage for the onset of diabetes.

All these conditions, incidentally, increase the person's cravings for sugar and processed foods. These substances stimulate the spleen and jog it into functioning at a higher rate for short periods. However, because the organ is weak, tired, and bloated, it cannot sustain normal function and lapses into lethargy once again. This only stimulates another round of cravings for sugar and processed foods. When it comes to overcoming cravings and addictive behavior, healing the pancreas and spleen is essential.

Healing the Spleen and Pancreas

Here are the key elements to healing the spleen. First, eat vegetables and grains that provide mild sweetness. The taste that gently supports and stimulates the spleen is the mild sweetness that comes from certain vegetables, especially the sweet squashes, such as the following:

Squash:

- Acorn squash
- Butternut squash
- Delicata squash
- Hokkaido pumpkin
- Hubbard squash
- Kabocha
- Pumpkin
- Yellow squash
- Zucchini

Other vegetables, including:

- Carrots
- Onions
- Rutabaga
- Shiitake mushrooms
- Turnips

Certain grains, including:

- Millet
- Sweet corn

Green and leafy vegetables, including:

- Collard greens
- Chard

Alkalizing foods, including:

Miso soup (see Chapter 10)
Soups and broths made with tamari and shoyu
Pickles

Certain fruits and nuts, including:

Red apple
Cantaloupe
Figs
Honeydew melon
Sweet orange
Raisins
Tangerine
Almonds

These foods heal the spleen by charging it with a gentle chi, or life energy. They also balance the blood's pH, and thereby balance the organ's acid-alkaline state, as well. In the process, they drastically reduce cravings and promote the healing of the organ.

Other Behaviors That Heal the Spleen

There are numerous other ways to heal the spleen and pancreas. The first of these is exercise. When you exercise, your spleen contracts in order to send more blood into the general circulation. More blood means more oxygen going to cells. But when you are resting, such as when you are sitting down, the body doesn't need as much blood. In that case, the spleen expands and takes up more blood into the organ.

A lack of exercise makes the spleen swollen, lethargic, and deficient. Not surprisingly, deficient spleen is associated with overweight—the proverbial couch potato. But when you exercise, the spleen contracts and is exercised itself. This restores the spleen's capacity to expand and contract, and its overall vitality. Exercise gets the spleen in shape again, thus making it more balanced, vital, and alive.

Another very effective way is to chew your food thoroughly to release a maximum amount of saliva. Saliva is slightly alkaline. Because the spleen has difficulty dealing with acidic foods, saliva acts as a balancing agent for the spleen. In general, saliva and all slightly alkaline foods are tonifying and strengthening to the spleen. Thus, the Chinese maintain that saliva is intimately connected with the spleen.

The condition of the spleen will sometimes affect how much we salivate. Often, when we suffer from indigestion, the salivary glands secrete more saliva, which in turn has a medicinal effect on the spleen. Those with chronically dry mouths, on the other hand, may suffer from deficient spleens. Chewing and salivating make food more accessible to our digestion, but at the same time promote the strength and vitality of the spleen and pancreas.

Acupuncture can promote an increase of chi, or life force, to the spleen and promote healing, as well.

That is the consequence of eating these foods every day: to eliminate all strong cravings and free us to enjoy a health-promoting diet and lifestyle. That is part of the purpose of the Energy Balance Program, and what it consistently accomplishes.

Healthy people are not possessed by cravings for ice cream, cheesecake, or candy. On the other hand, healthy people have the freedom to enjoy all kinds of foods, including an occasional piece of cheesecake or some ice cream, if they want such foods. But therein lies the difference between health and addiction: in health, we have the freedom of choice.

If you are having trouble losing weight or restoring your health, include more foods that promote the health of the spleen, as well as the body type that most closely fits you. See an acupuncturist, Chinese medical practitioner, or specialist in Ayurvedic medicine. Meanwhile, use the Energy Balance Diet as the basis for your way of eating. These practices, along with the Energy Balance Program, can guide you to greater health and the optimal body weight that you have been striving to achieve.

chapter 8

Overcoming Hidden Barriers

Here are just a few common statements made today by people about diet, health, and weight loss:

> "I avoid carbohydrates because they'll make me fat."
>
> "I eat lots of protein because protein will make me thin."
>
> "I eat lots of dairy foods, like milk and cheese, because they have calcium and protect my bones."

Not only is every one of these statements incorrect, but they are among the beliefs that keep people overweight and sick. Still, they have become so much a part of the public consciousness that they are now accepted as facts. Most people listen to statements like these, nod their heads knowingly, and then follow the advice without ever questioning if those statements are true. Ironically, people continue to believe in them, even as we in North America get sicker and heavier.

Has it occurred to anyone that if these statement were true, the Japanese would be the sickest and the fattest people on earth? The traditional Japanese diet is rich in carbohydrates, low in protein, and completely absent of dairy products. Yet the Japanese are healthy and lean, have low rates of cancer and heart disease, and have virtually no osteoporosis. According to the World

Health Organization, the Japanese are the longest-living people on earth, with an average life span of more than 81 years. But when they move to the United States and adopt an American diet, they become overweight and develop cancers of the breast, prostate, and colon, along with heart disease and osteoporosis.

Nevertheless, despite all the evidence to the contrary, people go right on believing that high-protein, low-carb, and high-dairy consumption is good for them. The question we have to ask ourselves is this: How can we hope to regain our health and achieve optimal weight if we are going about it in all the wrong ways?

When I ask people what is the biggest stumbling block they face in their struggle to lose weight or regain their health, most blame themselves. They see themselves as the reason they are failing in their health programs. Guilt blinds people from seeing the real reason why they are unable to achieve better health and optimal weight. That reason is the abundance of misinformation and pseudo-science being published about diet and health today. People are confused. Even well-read, intelligent people are confused because so much of the information contradicts itself. All this prevents people from knowing what to eat to lose weight and accomplish their health goals.

All of this reminds me of the old Sufi story about the man who lost his keys in his backyard but insisted on looking for them on a nearby street corner. When someone asked him why he was looking for his keys on the street corner instead of his backyard, he answered, "Because the light is better here." As long as we're looking for good health and optimal weight in the wrong places, we're never going to find it.

Whenever I say this in public lectures, people are invariably shocked. "How is it possible that we've been so misled?" they ask.

Food Guide Pyramid

That is a very good—and very important—question. And with that, I bring out the "Food Guide Pyramid, a Guide to Daily Food Choices," created by the U.S. Department of Agriculture (USDA).

The Food Guide Pyramid is the U.S. government's official guide to healthy eating and weight loss. It's supposed to represent what science knows about diet, health, and weight loss. It's supposed to guide you on what to eat every day so that you can become healthy and lean. The

Pyramid is used by every government-sanctioned health agency, including the American Dietetics Association, who are the people who determine what your children eat at school, and the food you are given in hospitals and other public institutions. It's the model for what children are taught in public schools about good eating habits. And finally, the Pyramid is also the basis for the advice given by dietitians or nutritionists. The Food Guide Pyramid, therefore, is the basis for much of our nutrition education.

Unfortunately, the Food Guide Pyramid does not represent what science knows about diet, health, and weight loss. It is loaded with misinformation that was deliberately put there to pacify the food industries. But what is especially disturbing is that the misinformation is responsible, in part, for high-protein diets, and such disorders as overweight, osteoporosis, heart disease, and many other serious diseases that people suffer from today.

"If you are really committed to finding your own healthy way of eating, getting your health back, and losing weight," I tell people, "you must become so sophisticated that you can see how people in corporations and in government are manipulating information to create confusion over diet, nutrition, health, and weight loss today."

Let's turn to the Food Guide Pyramid to show you what I mean.

Making Good Choices

The Food Guide Pyramid identifies six food groups and illustrates how much of each food group should be included in your diet. The Pyramid represents a reverse hierarchy, meaning that you are urged to eat more of the foods located at the bottom of the Pyramid than at the top. As you progress upward on the pyramid, you should eat smaller and smaller amounts of the foods listed.

The foods at the bottom of the Pyramid are the grains, or the carbohydrate category. The USDA urges us to eat between 6 and 11 servings of these foods per day. The next level up shows two different groups of foods, the vegetables, of which you should eat between three and five servings, and the fruits, of which you should eat two to four servings. The level above contains two more groups, the dairy products on one side and the protein sources on the other. Two to three servings of each should be eaten daily. Above that are the fats, oils, and sweets, which should only be eaten sparingly (whatever that means).

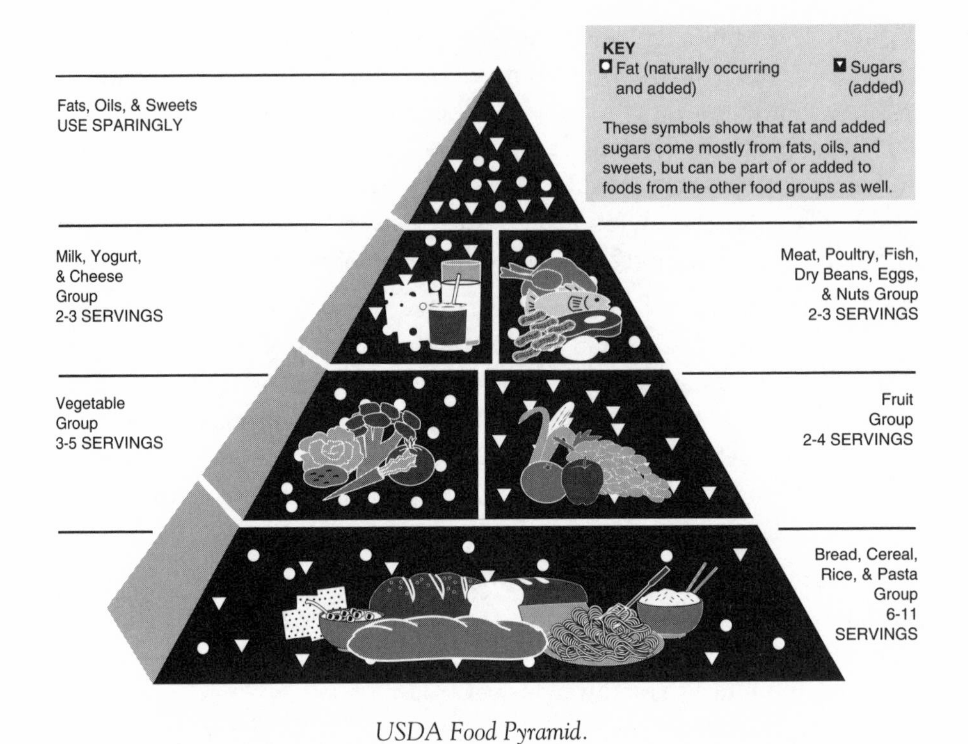

USDA Food Pyramid.

(Source: U.S. Department of Agriculture–U.S. Department of Health and Human Services)

In fact, this is not the original Pyramid, but a revised edition of it. The first edition of the Pyramid, which was to be called "The Eating Right Pyramid," was produced in 1991, but the meat and dairy industries blocked the publication of that guide because they claimed that the Eating Right Pyramid "stigmatized" (their word) dairy and meat.

The reason the dairy and meat industries felt "stigmatized" was because the USDA placed these foods in the "eat less" sections of the original Pyramid, writes Marion Nestle, Ph.D., in her book *Food Politics, How the Food Industry Influences Nutrition and Health* (University of California Press, 2002). Dr. Nestle, who is professor and chair of the Department of Nutrition and Food Studies at New York University, reports that the USDA then pulled its Eat Right Pyramid and spent nearly a million dollars trying to find ways to appease the meat and dairy industries. A year later, the department came up with a new Pyramid and a new name. The new Food Guide Pyramid, with its altered recommendations, now "sufficiently appeased the meat and dairy industries," Dr. Nestle reports, and therefore could be published.

As she says in her book, "Although the Pyramid can be viewed as just another example of government's caving in to industry pressure," it reveals "the conflict of interest created by the USDA's dual mandates to protect agriculture and to advise the public about diet and health, and the influence of lobbyists in this area ..." Guess who wins when the USDA is forced to confront that conflict of interest? You guessed it.

In fact, there's a long history of the government turning its back on sound dietary advice when it threatens the meat and dairy industries. In 1977, the U.S. Senate Select Committee on Nutrition and Human Needs published its now famous "Dietary Goals for the United States," the first government report that clearly showed that red meat, dairy products, eggs, and processed foods were causing most serious illnesses and overweight. The original Dietary Goals stated plainly that Americans should "decrease consumption of meat," a recommendation that caused the meat industry to suffer a proverbial heart attack. The president of the National Cattlemen's Association, Wray Finney, told Senator Robert Dole, the Republican Senator from Kansas, that the sentence had to be deleted from the report.

In her book, Dr. Nestle reports a conversation between Dole and Finney that changed the report significantly. Dole asked Finney if it would be okay with the National Cattlemen if the report said, "'Increase consumption of lean meat?' Would that taste better to you?" Dole asked.

"Decrease is a bad word, Senator," replied Finney. The sentence never appeared again in Dietary Goals.

We should ask ourselves, for whom is *decrease* a bad word? For people who are overweight? For those who are sick, or on the brink of becoming ill? Ask yourself, is decrease a bad word for you? Or is it the right advice for you at this moment in your life?

Is there any doubt in your mind that people should decrease consumption of meat and meat products? The fact is that an abundance of science clearly shows that meat is causing an array of serious illnesses, including heart disease and cancer. It would be good for everyone in the Western world to reduce meat intake, which would result in a dramatic decrease in serious illness and overweight, not to mention health-care costs. Unfortunately, the government is not allowed to tell you that, because lobbyists, who contribute huge sums of money to election campaigns, are able to muzzle our elected representatives.

It's worth noting that most of the nutrition research done today is paid for by tax dollars doled out by the National Institutes of Health, a federal group of scientific agencies under the direction of the Department of Health and Human Services. Unfortunately, our government cannot tell us what our tax dollars are discovering, because it might offend the meat and dairy industries.

This is more than a history lesson. The conflict of interest that Congress and our federal agencies face directly affects our most widely held beliefs about food, and therefore determines to a great extent whether or not you will be successful in your health goals. Let me show you how.

Is Bread Really the Same as Brown Rice?

According to the Food Guide Pyramid, the foods that should dominate your diet are the carbohydrate-rich foods, namely, bread, cereal, rice, and pasta. It says that, ideally, you should eat 6 to 11 servings of these foods per day.

First, it should be pointed out that the Pyramid makes no distinction between processed and whole grains. You have to hunt through the accompanying literature to find any reference to the difference between processed foods and whole grains. According to the Pyramid, white bread is as nutritious as whole-wheat sourdough, and white rice as nutritious as brown rice. It is also implied that whole-grain breads, such as wheat bread, are considered the equal of whole-grain foods, such as brown rice, millet, or barley.

Therefore, you would be complying with the Pyramid if you ate 6 to 11 servings of bread each day. Of course, you could also have "three or four crackers," according to the literature that accompanies the Pyramid. The crackers would represent another serving of grain products.

Bread provides more than 1,200 calories per pound. Oyster crackers provide as many as 1,900 calories per pound; saline crackers, such as Saltines, provide more than 1,600. Because little apparent distinction is made between processed versus whole grains, you might easily believe that you are doing the right thing by choosing, say, pretzels—they are low in fat, after all—or granola bars. Pretzels provide 1,770 calories a pound, a granola bar 2,300.

On the other hand, unprocessed grains are in another class entirely. Brown rice provides about 500 calories per pound, and millet, quinoa, and amaranth considerably less. Even pasta, which gets a very bad rap today, is a healthier and less calorically dense choice. A pound of pasta contains about 400 calories per pound. A large bowl of pasta, which is still considerably less than a pound, provides only 160 calories. If you add lots of vegetables to that bowl, you'll fill up on even fewer calories.

Not only are processed foods richer in calories, but they are also rapidly absorbed. That means that it keeps insulin levels up and promotes weight retention and weight gain. Needless to say, processed foods are also lower in many nutrients that are present in whole grains.

Now let's say that you want to improve your health and lower your weight. Obviously, the distinction between processed foods and whole foods is an important one to you. But when you are taught that all carbohydrates are alike, and all carbs will add pounds to your body, guess which foods you are likely to avoid, and which ones you are likely to eat? That's right, you'll avoid the carbs and choose animal foods—namely meat, eggs, chicken, fish, and dairy products. Our misconceptions about carbohydrates form the foundation for high-protein diets, and our over-reliance on animal protein. And these foods are having a tremendous impact on our health and our weight.

One of the problems facing us today is the word "carbohydrate," which is too general and too vague for us to know if the food is processed or whole. All carbohydrates are not created equal. We should throw out that word, or at the very least ask ourselves a basic question whenever we hear the word carbohydrates: Are they processed or whole?

What the Food Guide Pyramid should be telling everyone is that the recommended food is unprocessed whole grains. There should be three categories of grain products, which would look something like this.

Preferred	Use in Moderation	Not Preferred
Whole grains, such as:	(Use in moderation to promote weight loss)	Bagels
Amaranth	Whole-grain pasta	Chips
Barley	Buckwheat (soba) pasta	Cookies

continues

continued

Preferred	Use in Moderation	Not Preferred
Brown rice	Udon (sifted wheat) noodles	Crackers
Buckwheat	Semolina pasta	Granola bars
Millet	Rice noodles	Muffins
Oats	White rice	Pastries
Quinoa	Whole-grain bread	White bread
Teff		

Every civilization on earth has been founded on whole grains—not simple sugars or processed foods. When I tell people the difference between whole versus processed grains, they immediately understand and feel relieved. "Oh, so I can eat carbohydrate foods," they say to me. "Yes," I say, "as long as you make them whole and unprocessed."

Vegetables and Fruit: The Benefits

The Pyramid recommends that you eat three to five servings of vegetables a day. It recommends two servings of fruit a day, as well. Vegetables, grains, beans, and fruit are the keys to optimal nutrition, weight loss, more beautiful skin, and a more healthful and relaxed body.

Three servings of vegetables are not enough. If you eat only three servings, it means you're probably filling up on animal and processed foods—which means you're getting lots of fat and calories.

Here's the basic axiom in dieting today:

> You're either eating vegetables, or you're eating animal and processed foods.

There's very little in between with most people. And if you're eating lots of the latter, you're getting fatter and sicker. Five servings of vegetables is the minimum. If you want to lose weight, I recommend that you try to eat seven servings per day.

The problem for most people is that they wouldn't know how to eat five or seven servings of vegetables. When asked to think about vegetables, most people think of french fries, ketchup, potato chips, and salad.

I have nothing against salad, but it is not a particularly nutritious vegetable source. Most salads are made of iceberg lettuce, which is basically fiber and water. It has virtually no nutritional value.

Our relationship with vegetables reflects how far removed we are from nature. Vegetables are no longer familiar to us. The food that nature used to help us survive and evolve is now foreign to most Westerners. If it doesn't come in a package, like frozen peas and corn, it no longer makes sense to us. People no longer think of cooked greens with a delicious sauce, onions and carrots sautéed in olive oil, sweet squash, or a root stew in miso or tamari broth. Cooked fresh vegetables provide lightness to the body. They are extremely low in calories, but rich in nutrition and fiber. They improve circulation and elimination, and provide a rich supply of vitamins, minerals, and immune-boosting antioxidants, carotenoids, and phytochemicals. Vegetables are the basis for good health.

There are literally thousands of edible plants, and hundreds of green and leafy vegetables. But ask yourself: How many people regularly eat even the most familiar green and leafies, such as broccoli, cabbage, collard greens, kale, mustard greens, and watercress? How many of us regularly eat the most familiar roots, such as carrots, parsnips, rutabagas, and turnips, or the most familiar round vegetables, such as onions or squash? The irony is that these foods are easy to prepare, and they require very little time. Most can be cleaned, cut, and steamed in fewer than 10 minutes. And they are the keys to good health and optimal weight. The very answers to our most intractable health problems are growing right under our noses and being ignored, because they are no longer familiar to our experience.

Here's a fundamental truth about achieving better health, optimal weight, and personal transformation:

> If you are eating an abundance of vegetables every day, you are almost guaranteed to change your life for the better.

But if you are eating only a few vegetables each day, you're very likely going toward greater sickness, greater weight, and greater problems. It's that simple. Become a vegetable eater and change your life. Chapter 10 will give you some enticing ways to cook veggies.

Are Beans Really the Same as Meat, Poultry, or Fish?

The Pyramid urges you to eat two to three servings a day of "Meat, poultry, fish, dry beans, eggs, and nuts." Despite what the Pyramid leads us to believe, these are not interchangeable foods. They have vastly different effects on your health, as scientists have been telling us for the last 30 years. Red meat and chicken contain significant amounts of fat, especially saturated fat, which has been shown to promote heart disease, cancer, adult-onset diabetes, and other serious illnesses.

Most of the red meats used in the United States contain small but significant traces of hormones, which are used to fatten animals. Researchers are increasingly concerned that those hormones are among the reasons that girls are coming to maturity, and experiencing early menarche, or first menstrual period, at younger and younger ages. Studies have shown that the younger a woman experiences her first period, the greater her risk of contracting breast cancer later in life.

The fat in red meat also serves as a repository for pesticides, herbicides, and radioactive particles. Many of these substances have estrogenic effects, as well, which means they increase the risk of breast, ovarian, and uterine cancers. Similar concerns exist for chicken, which are also exposed to hormones and are richer in fat.

The fat in red meat makes it rich in calories, as well. A pound of top sirloin provides 870 calories per pound; top round, 950; veal, 980; prime round, 975; and a rib-eye steak, 1,020. Chicken, without the skin, provides 750 calories per pound.

The fat content of fish is very different from that of red meat in both quantity and quality. White fish, such as cod and halibut, provide only tiny amounts of saturated fat (usually less than one percent of their calories), but are abundant in omega-3 polyunsaturated fats, which boost your immune and cancer-fighting systems. White fish, typically, are low in fat and calories. Orange roughy, for example, provides 400 calories per pound; cod, 480; and halibut, 520.

Some fish, especially those that swim in dirty waters, have tiny to significant amounts of mercury, among the most deadly poisons on earth. Not all fish contain mercury, however. Most white fish are essentially free of this powerful toxin. A good fish market tests its fish and can tell you if they are free of mercury toxicity. Ask your fishmonger if his fish have been tested for mercury, and if so what are the results.

The protein in animal foods has been shown to create an array of health issues, as I showed in Chapter 6. Animal protein is much more difficult for the human body to metabolize and has an array of by-products, such as ammonia and uric acid. Vegetable proteins, which come from beans and grains, are much easier for the body to metabolize and do not result in the same by-products that animal proteins do.

Once we get out of the animal sources of protein, we enter a whole other world of nutrition. Beans provide a rich array of nutrients, antioxidants, and isoflavones, including genistein and daidzien, two powerful cancer fighters. Dry beans provide anywhere from 400 to 600 calories per pound, both levels well within the ranges that will promote weight loss.

As for nuts, almonds and walnuts can lower LDL cholesterol and protect your heart. Almonds also contain significant amounts of vitamin E; walnuts have selenium, and both provide heart-healthy magnesium. These two antioxidants provide numerous health benefits. But nuts are richer in fat—albeit monounsaturated fat. Still, even monounsaturated fat can raise your weight. Nuts are good for us, but we should eat them in moderation.

By joining meat, eggs, poultry, fish, beans, and nuts, the Pyramid has combined foods that have vastly different effects on our health and weight. You should know this if you are at all concerned about either.

Do We Need Two to Three Servings of Milk, Yogurt, and Cheese?

The Pyramid urges us to eat two to three servings a day of milk, yogurt, and cheese, which has become almost universally accepted wisdom. Dairy foods are not considered an option, but are regarded as essential, especially in light of the osteoporosis epidemic. Yet most non-Caucasian people cannot digest dairy products after they are weaned because they lose the enzyme lactase, which is needed to digest the sugar in milk, known as lactose. Those who lose lactase include all Asians, most people of African descent, and many Hispanics.

Many people of color argue that the U.S. government's urging that everyone drink milk is racist. The Black Caucus stated that the USDA is guilty of "a consistent racial bias." When you consider that the U.S. government is urging all its people to eat dairy foods, including Asian, African American, and Hispanic children, even when it knows that many of these children will suffer a variety of painful digestive disorders as a consequence, you begin to see how far the USDA will go to appease the National Dairy Council.

As I have already shown in Chapters 3 and 6, a growing body of scientific research has linked dairy foods to an array of serious illnesses, including autoimmune disorders, juvenile diabetes, heart disease, allergies, digestive disorders, leaky gut syndrome, iron-deficiency anemia, and cancers of the ovaries and uterus. In addition, dairy foods have consistently turned up trace residues of growth hormones, antibiotics, and pesticides. As with meat products, many scientists are concerned that the estrogenic effects of these substances may increase the risk of cancer.

You can look high and low in the USDA guidelines and you will not find any of this information, including the fact that most non-Caucasians should avoid milk products.

"How dairy foods came to be considered essential despite their high content of fat, saturated fat, and lactose is a topic of considerable historical interest," writes Dr. Nestle. "As it turns out, nutritionists have collaborated with dairy lobbies to promote the nutritional value of dairy products since the early years of the twentieth century. Recently, however, some scientists have raised doubts about whether dairy foods confer special health benefits. In addition to concerns about lactose intolerance, some question the conventional wisdom that dairy foods protect against osteoporosis, or, for that matter, accomplish *any* public health goals. Others suggest that the hormones, growth factors, and allergenic proteins in dairy foods end up doing more harm than good." [Dr. Nestle's emphasis.]

Historically, the leading source of diet and nutritional information in America has been the National Dairy Council. When the USDA's Pyramid was finally published, the Dairy Council created its own pyramid, which included milk products in every food category. You might argue that the dairy industry has a right to do that, since it wants to promote its own products. But in this case, the product has serious adverse effects on health and weight, especially for certain ethnicities. That information is not widely disseminated. Indeed, the only information many people get about milk products is from the National Dairy Council. To an increasing number of scientists and medical doctors, including the Physician's Committee for Responsible Medicine (PCRM), the actions of the milk industry are reminiscent of the cigarette lobby, which also promoted a product that caused serious illness, but kept that information from public knowledge. The PCRM has assiduously researched the health effects of dairy products and describes those effects in detail on its website (PCRM.com).

When it comes to promoting your health and optimal weight, you have to recognize that milk consumption bears special scrutiny and personal experimentation. One of the things I have seen is that when people stop eating dairy foods, two things happen: first, many allergies and chronic diseases clear; second, people lose weight, often without making any other adjustment in their diets. Milk is not essential to health. For many people, it can be highly injurious—for some even life-threatening.

It's important to recognize that infants do not need cow's milk. They require breast milk, which confers a seemingly endless supply of benefits. Among those benefits is breast milk's unique capacity to strengthen the human immune system, and to make it stronger for the person's entire life. Breast milk also increases the body's ability to metabolize fats and cholesterol. Numerous studies have shown that breast-fed babies grow up to have higher IQ scores than those who were not breast fed. Once weaned, children do not require milk products in order to maintain health and strong bones. On the contrary, most children around the world do not consume milk products after weaning. As long as they eat a steady diet of plant foods, their bones and teeth remain strong and their calcium levels normal.

Do Most Fats, Oils, and Sweets Come Separately?

Among the more interesting features of the Pyramid is that it provides a separate category for fats, oils, and sweets, as if these were foods unto themselves. Most of the fat people consume come as part of food, namely animal, dairy, and processed foods. The same is true of oils, which are most prevalent in fried and processed foods. As for sweets, they appear most often in pastries, desserts, and soft drinks. Yes, sugar can be consumed separately, as can oil. But the overwhelming preponderance of fat, oil, and sugars are consumed as part of foods.

The Pyramid urges us to eat these substances sparingly. But if you follow that advice—and that's an essential step in health promotion—you will have to reduce your consumption of animal, dairy, processed, and fried foods, along with sugary desserts and soft drinks. When we face what it means to cut back on those foods, we must ask ourselves, What's left? The answer is a diet composed chiefly of whole grains, vegetables, beans, and fruit. That diet is made up primarily of carbohydrates—complex carbohydrates from whole, unprocessed foods. Finally, we get to the advice that will really work!

The Energy Balance Diet Addresses All the Ills

People make changes in their diets for many reasons: They want to look better, lose weight, overcome disease, or develop themselves in new ways. Whatever your reason may be, it's a good one, as far as I am concerned. That desire to change places you on a seeker's path. But like any seeker, you must sort through all the information and misinformation to find the truth.

As you search for answers, remember one of the basic tenets of capitalism: *caveat emptor*, or buyer beware. It means that in the end, you are responsible for your choices, for the goods that you buy, for the information that you accept as true, for the food that you put in your mouth. The marketplace is filled with programs and promises. There is an incredible amount of money at stake. Powerful industries can influence even our most fundamental institutions.

Look below the surface of things and examine the evidence that may or may not support those promises. Once you decide to use diet to improve your health and your appearance, you are suddenly thrown into a maelstrom of voices, all claiming to have the answer to your problem. You have to ask your own body what makes it feel better or worse. You have to trust in your own experience: Is this way of eating making me feel and look better, is it working, or is it just another set of empty promises?

It can be helpful to remember that the standard Western diet, loaded with fat, cholesterol, processed flours, sugar, and artificial ingredients, is toxic to the human body. It causes a wide array of problems, everything from heart disease and cancer to obesity and depression.

The answer to these problems is to go back to the way of eating that we evolved on, the way of eating that helped to shape the human form and gave it such awesome regenerative powers. That way of eating is essentially a plant-based diet, supplemented by low to moderate amounts of animal foods. Such a regimen can reverse most of the problems that the modern diet has caused. It can give you back your health, your optimal weight, the stability of your emotions, and the clarity of your mind. These gifts are foundations of an effective, successful, and happy life.

chapter 9

Meal Plans That Heal

Following are three weeks' worth of menus for breakfasts, lunches, and dinners on the Energy Balance Diet. In Chapter 10, I have provided more than 120 recipes for healthful meals. Virtually every entrée listed in this chapter has a corresponding recipe in Chapter 10. Also, there are more recipes than meals provided below, so be sure to try all the recipes in Chapter 10 to provide new and creative meals for yourself, your loved ones, and your friends. Chapter 10 also provides a long list of healthful desserts. In Chapter 4, I have provided a long list of snacks, which can be eaten between meals.

I have included numerous suggestions for fish dishes and organically raised chicken and red meat. By organic and range-fed, I mean animals that have not been subjected to hormones, antibiotics, or other synthetic chemicals to promote growth or greater milk production. Organically raised animals have also been allowed to roam freely and eat healthful feed. I recommend that you limit the size of your portions of animal food to 3½ ounces.

As I have said throughout the book, my first choice for animal food dishes is fish, but for people who desire chicken or red meat, I recommend that you choose the highest quality possible. I, for one, enjoy red meat periodically, especially beef, and choose organically raised beef whenever I eat it. In the menus, I have listed a fish entrée, but have suggested chicken or red meat as an alternate to fish. In other words, you don't have to have fish every time I suggest fish, but can alternate chicken or red meat as you prefer. Again, choose organic quality whenever possible and limit serving size.

The concerns for quality extend to fish, as well, especially for the presence of mercury. Many fishmongers today test their fish for mercury toxicity, as well as the presence of hormones and antibiotics. Salmon, for example, are typically raised on farms and are often subjected to the same kinds of steroids and antibiotics that are injected into livestock. I strongly urge you to buy only organic fish and fish that have been harvested from clean waters. It's actually a lot easier than you might think to get mercury-free fish and wild Alaskan salmon that has not been subjected to synthetic growth stimulants and antibiotics. Many salmon growers are now producing organic salmon, as well. It's common practice at many natural foods markets today to provide organic, mercury-free fish and organic chicken and red meats. My point in all of this is to urge you to choose quality when buying fish, chicken, and red meat.

Most of the foods on the Energy Balance Diet can be prepared in fewer than 30 minutes. The foods are also very easy to prepare and delicious. Not only will they help you achieve your health and weight goals, but you'll enjoy eating on this program.

Remember to drink water throughout the day. Even though this is not listed in the menu guide, it's important. Keep a bottle of spring water on your desk and sip it throughout the day. I have listed tea in the menu plans. Tea can be noncaffeinated herbal tea, or black or green tea, depending on what you desire. Both black and green teas have caffeine, of course, but they are also loaded with antioxidants. Tea is a healthful substitute for the caffeine kick in coffee. Please do your best to avoid coffee. Coffee is an extreme food and will create severe cravings.

After three weeks on the Energy Balance Diet, you will very likely experience a diminution in all your existing physical symptoms, a reduction in weight, and a significant decline in your cravings for extreme foods. When that happens, you'll know you are on your way to better health, optimal weight, and a new relationship with yourself.

Week I

Monday

Breakfast:

Oatmeal and raisins (cooked together) (If desired, add any of the following: freshly ground flax seeds, or cracked and roasted flax seeds, wheat germ, and, if a sweet taste is desired, a small amount of rice syrup.)

Tea, grain coffee, or water

Lunch:

Fish sandwich (Steam white fish 5 to 10 minutes; place on spelt bread, or rye; add mustard, lettuce, tomato.)

Salad with olive oil and lemon dressing

Tea, grain coffee, or water

Dinner:

Shiitake miso soup (See Chapter 10.)

Basic brown rice

Baked aduki beans

Sautéed carrot, onion, and arame

Steamed kale

Beverage of choice

Tuesday

Breakfast:

Morning glory with oats (See Chapter 10.)

Tea, grain coffee, or water

Lunch:

Simple soba noodles

Black beans with tomato and garlic

Salad with dressing
Tea, grain coffee, or water

Dinner:

Mighty miso soup
Kamut brown rice
Broiled salmon or 3-oz. serving of broiled lean beef
Spring nishimi
Steamed kale
Beverage of choice

Wednesday

Breakfast:

Scrambled tofu
Toast, rye, spelt, or oat bread
Tea or grain coffee

Lunch:

Roasted chestnut/lentil/shiitake salad
Steamed greens
Tea, grain coffee, or water

Dinner:

Creamy broccoli soup
Buckwheat with carrot and arame
Mung bean paté
Roasted rutabaga with celery root and tamari almond
Pressed salad
Beverage of choice

Thursday

Breakfast:

Creamy kasha (See Chapter 10.)

Tea, grain coffee, or water

Lunch:

Fish sandwich

Salad with dressing

Tea, grain coffee, or water

Dinner:

Mighty miso soup

Gypsy singing rice salad

Red lentil burgers

Sautéed carrot, onion, and arame

Steamed kale

Beverage of choice

Friday

Breakfast:

Muesli

Tea, grain coffee, or water

Lunch:

Fried rice with shrimp and broccoli

Pressed salad

Tea, grain coffee, or water

Dinner:

Hot borscht (beet soup)

Millet/carrot/hiziki/burdock
Baked tofu
Steamed greens
Beverage of choice

Saturday

Breakfast:

Scrambled tofu
Tea, grain coffee, or water

Lunch:

Sesame-dulse brown rice
Spring nishimi
Sautéed greens
Minty green lentil-watercress salad
Tea, grain coffee, or water

Dinner:

Shiitake miso soup
Broiled chicken (Coat skin with olive oil and small amount of salt and pepper before broiling.)
Cold soba salad
Ginger-garlic broccoli and toasted sesame seeds
Steamed greens
Beverage of choice

Sunday

Breakfast:

Two-egg omelet with vegetables
Tea, grain coffee, or water

Lunch:

Basic brown rice

Parsley stir-fry topping

Boiled salad

Sweet sensation

Tea, grain coffee, or water

Dinner:

Creamy chestnut-squash soup

Basic quinoa

Red lentil burgers

Spring nishimi

Steamed kale

Beverage of choice

Week 2

Monday

Breakfast:

Cream of kasha

Tea, grain coffee, or water

Lunch:

Basic millet

Vegetable soup

Sautéed vegetables with tofu

Tea, grain coffee, or water

Dinner:

Creamy salmon soup

Golden rice

Bok choy stir-fry
Steamed daikon with black sesame seeds
Beverage of choice

Tuesday

Breakfast:

Morning glory with oats
Tea, grain coffee, or water

Lunch:

Simple soba noodles
Salad with dressing
Black beans with tomato and garlic
Tea, grain coffee, or water

Dinner:

Mighty miso soup
Kamut brown rice
Broiled Alaskan salmon or 3-oz. serving of chicken or lean beef
Spring nishimi
Steamed kale
Beverage of choice

Wednesday

Breakfast:

Oatmeal with raisins
Tea, grain coffee, or water

Lunch:

Black beans on whole-wheat tortilla with lettuce, tomato, and onion

Tea, grain coffee, or water

Dinner:

Carrot ginger soup

Very easy fried rice

Marinated tofu stir-fry

Baked rosemary-caraway sweet potato

Blanched edamame

Steamed greens

Beverage of choice

Thursday

Breakfast:

Scrambled tofu

Tea, grain coffee, or water

Lunch:

Broiled whitefish

Salad with dressing

Steamed Chinese Napa cabbage and shiitake mushrooms (Could substitute bok choy for Napa cabbage.)

Tea, grain coffee, or water

Dinner:

Aduki squash stew

Basic brown rice

Little mermaid salad

Boiled salad

Sautéed greens

Beverage of choice

Friday

Breakfast:

Muesli

Tea, grain coffee, or water

Lunch:

Pinto beans on soft-shell tortilla with lettuce, tomato, and onions

Tea, grain coffee, or water

Dinner:

Creamy broccoli soup

Winter squash and millet

Marinated tofu stir-fry

Sautéed carrot, onion, and arame

Steamed greens

Beverage of choice

Saturday

Breakfast:

Quick-cooking seven-grain cereal (Add raisins, strawberries, blueberries, sliced banana.)

Tea, grain coffee, or water

Lunch:

Very easy fried rice

Scrambled tofu

Broccoli with garlic and ginger

Tea, grain coffee, or water

Dinner:

Split pea soup

Millet with quinoa

Sautéed greens

Broiled salmon, chicken, or lean beef

Steamed daikon

Beverage of choice

Sunday

Breakfast:

Morning glory with oats

Tea, grain coffee, or water

Lunch:

Buckwheat with carrot and arame

Sautéed cabbage

Tea, grain coffee, or water

Dinner:

Carrot ginger soup

Millet smashed potato

Basic chickpeas

Steamed collards

Beverage of choice

Week 3

Monday

Breakfast:

Cream of kasha
Tea, grain coffee, or water

Lunch:

Fried rice with shrimp and vegetables
Salad with dressing
Tea, grain coffee, or water

Dinner:

Shiitake miso soup
Executive rice
Black bean salad
Sautéed cabbage
Garlic steamed string beans
Beverage of choice

Tuesday

Breakfast:

Morning glory with oats
Tea, grain coffee, or water

Lunch:

Whitefish salad sandwich
Vegetable soup
Tea, grain coffee, or water

Dinner:

Shiitake miso soup

Long-grain brown rice with chestnuts

Beet, carrot, parsnips fennel extravaganza

Marinated tofu stir-fry

Steamed leafy greens

Beverage of choice

Wednesday

Breakfast:

Oatmeal with raisins (If desired, add any of the following: fresh ground flax seeds, or cracked and roasted flax seeds, wheat germ. Add rice syrup if sweet taste is desired.)

Tea, grain coffee, or water

Lunch:

Hummus sandwich on whole-grain bread (rye, rice, oat, or spelt bread) with lettuce, tomato, and onion

Tea, grain coffee, or water

Dinner:

Hot borscht

Broiled chicken or white fish with wild rice

Carrot/burdock power meal

Steamed greens

Beverage of choice

Thursday

Breakfast:

Oatmeal and raisins (If desired, add any of the following: cracked and roasted flax seeds, wheat germ, and rice syrup for sweetening.)

Tea, grain coffee, or water

Lunch:

Vegetable soup

Scrambled tofu

Steamed kale

Tea, grain coffee, or water

Dinner:

Miso shiitake soup

Millet with sunflower seeds

Roasted kabocha squash

Minty green lentil/watercress salad

Basic chickpeas

Beverage of choice

Friday

Breakfast:

Muesli

Tea, grain coffee, or water

Lunch:

Quinoa salad

Sautéed broccoli with tofu

Tea, grain coffee, or water

Dinner:

Green millet

Broiled salmon or chicken

Steamed daikon with black sesame seeds

Garlic steamed string beans

Salad with dressing

Beverage of choice

Saturday

Breakfast:

Dry boxed cereal with apple juice or soymilk—add cut-up fruit

Tea, grain coffee, or water

Lunch:

Tuna sandwich with lettuce, tomato, and onion on rye, rice, or oat bread

Tea, grain coffee, or water

Dinner:

Shiitake miso soup

Zen buckwheat

Carrots, beets, onions, and tofu

Steamed kale

Beverage of choice

Sunday

Breakfast:

Scrambled tofu

Tea, grain coffee, or water

Lunch:

Quinoa salad
Bok choy stir-fry
Tea, grain coffee, or water

Dinner:

Split pea soup
Broiled whitefish, chicken, or 3 oz. red meat
Pressed salad
Roasted rutabaga with celery root and tamari almond
Beverage of choice

Recipes for these and other exciting foods can be found in the next chapter. Enjoy.

chapter 10
Recipes

Following are more than 120 recipes that will guide you in the preparation of everything from soup to desserts. Also provided are easy-to-follow instructions for preparing these meals. The order is as follows:

- Simple grain recipes
- Vegetables
- Salad and vegetable dressings
- Beans
- Soups
- Tofu, tempeh, and natto
- Sea vegetables
- Fish
- Desserts

Most of the recipes that follow can be made in 30 minutes or less. The exceptions are some of the grains and beans, which can take longer. *Bon appétit!*

Simple Grain Recipes

Basic Brown Rice Pot

1 cup brown rice
2 cups water
Pinch sea salt

Presoak. Gently rinse rice. Presoaking will make the rice more digestible, but is not necessary. If desired, presoak for one hour.

Add water and a pinch of sea salt. Bring to boil. Cover.

Reduce heat to low. Simmer for 50 minutes if it's short-grain rice and 35 minutes if it's long-grain basmati rice. When it's done, pull from heat and let stand covered for 10 more minutes.

Fluff with fork and serve.

White Basmati Rice

1 cup white basmati rice
1½ cups water
Pinch sea salt

Wash and drain rice.

Add water. Add salt.

Bring to boil. Cover and simmer for 30 to 35 minutes.

Do not touch rice during cooking, as it makes the rice mushy. When it's done, pull from heat and let steam covered for 10 more minutes.

Fluff with fork and serve.

Long-Grain Brown Rice with Chestnuts

½ cup dried chestnuts

1 cup rice

2½ cups water

Pinch sea salt

Soak chestnuts for at least eight hours.

Wash and drain rice.

Cut soaked chestnuts into bite-size pieces.

Put rice in water. Add salt. Bring to a boil.

Add chestnuts.

Cover and simmer for 45 minutes. When finished, pull from heat and let stand covered for 10 more minutes, then stir the rice mixture so the top mixes with the rest of the pot.

Golden Rice

2 cups white basmati rice

4 cups water

½ tsp. turmeric

½ tsp. cumin seeds

¼ tsp. sea salt

Wash and drain rice.

Add water and spices. Mix gently before adding heat. Bring to a boil, reduce heat, and cover. Simmer for 35 minutes.

Remove from heat and tenderly fold rice into the spices. In this dish, spices tend to dry out rice, so add 1 teaspoon ghee or 1 tablespoon olive oil.

Let the beautiful aroma take you places! Enjoy!

Ghee

If ghee is kept covered, it will keep indefinitely without refrigeration.

1 lb. raw, unsalted, organic butter

Sterilize a storage jar, pan, and spoon by immersing them in boiling water. Cook the butter in a stainless-steel frying pan over moderate heat for about 15 minutes. Stir gently and continually to avoid burning. Allow the foam that surfaces to settle on the bottom of the pan. When the ghee begins to boil, it is done. Allow to cool and then pour ghee into a clean container. Make sure that the foam remains in the frying pan.

Executive Rice

2 cups short-grain brown rice
3½ cups water
4 inches kombu

Wash and drain rice. Place in pressure cooker with water and kombu. Start cooking for 20 minutes covered, on low heat, then turn to high until your pressure valve starts to dance. Lower heat to simmer for 45 minutes.

Do not expect to have your first trial perfect. Sometimes you need to figure out the right balance of water and rice. It is a delicious, nourishing dish and worth it to initiate yourself to the magic of pressure-cooked rice!

Kamut Brown Rice

1 cup long-grain brown rice
1 cup kamut
4 cups water
Pinch sea salt

Wash and drain rice and kamut separately. Kamut is a very hearty grain and takes longer to cook.

Add kamut to water, add salt, and bring to boil. Cover and simmer for 15 minutes.

Add rice and simmer for an additional 40 minutes covered. You might want to check on water after 20 minutes, if it's absorbed, add another ¼ cup to your pot. When done, pull aside and let stand covered. Such an exciting combination of these royal grains!

Optional: If you want to add extra texture to your dish, add toasted sesame oil or ghee to the serving.

Optional: To make this a festive party meal, sprinkle with finely chopped parsley, scallions, or basil. Listen for the wows!

Sesame Dulse Brown Rice

2 cups brown rice
4 cups water
½ cup toasted black sesame seeds
¼ cup dulse

Wash and drain rice. Add to water and bring to a boil.

Meanwhile, toast sesame seeds in a skillet over medium heat. Seeds are done when they release their full nutty flavor into the air.

Add dulse. If the dulse is not already ground up, you may have to tear it into small pieces, before adding it to the seeds.

When water is boiling with the rice, add the seeds and dulse. Mix, cover, and simmer for 45 minutes. One of those rainy-day dishes that makes you want to go out and puddle jump!

Very Easy Fried Rice

1 small onion, chopped
2 cloves garlic, shredded
1 medium carrot, diced
½ bunch scallions
Grated ginger
4 cups cooked long-grain brown rice
1 tsp. toasted sesame oil
2 TB. tamari soy sauce
2 TB. chopped parsley

Sauté onion in toasted sesame oil.

Shred garlic into onion.

Add carrots and sauté for four minutes.

Add scallions.

Shred ginger and add.

Sauté these for about four more minutes so flavors can melt into each other.

Add rice and sprinkle with water. Water gives extra steam to the dish.

Add tamari soy sauce.

Lower heat and cook for 10 more minutes, stirring occasionally. When ready, pull aside, garnish with chopped parsley, and serve. You have now mastered an ancient art.

Wild Wild Rice

1 cup wild rice

4 cups water

Pinch sea salt

Wash and drain rice. Add water.

Bring to a boil, adding salt. Turn heat to low, cover and simmer for 45 to 50 minutes. Your grain is ready when black seeds are opened up.

Mix and serve. You are eating pure eternity!

Gypsies' Singing Rice Salad

1 cup cooked brown basmati rice

1 cup cooked white basmati rice

1 red bell pepper

1 medium purple onion

2 stalks celery

½ cup white sesame seeds

½ cup toasted pumpkin seeds

2 TB. olive oil

¼ tsp. black pepper

1 cup parsley, chopped

Optional: ½ lemon and ⅓ cup chopped mint leaves

If you use leftover rice, put it in the steaming basket to reenergize it and steam for 10 minutes. Otherwise, prepare rice according to basic rice preparation method.

Chop bell pepper and onion.

Dice celery.

Toast sesame and pumpkin seeds. You can combine them as you toast them; make sure you don't let them get dark brown.

Combine all ingredients in a big bowl, adding olive oil, black pepper, and parsley.

Add zest by squeezing ½ a lemon and adding ⅓ cup chopped mint leaves.

This is an exciting dish to amaze your crowd—and a simple and perfect dish to create with leftover rice. *Olé!*

Basic Millet

1 cup millet
3 cups water
Pinch sea salt

Wash and drain millet. For a nutty flavor, dry-roast millet until it smells toasty but is not brown.

Add salt to water and bring to a boil. Cover and cook for 25 minutes.

Millet and a pillow have the same affect on humans; they allow you to rest in peace after having a meal.

Millet with Roasted Sunflower Seeds

1 cup millet
½ cup toasted sunflower seeds
3 cups water
Pinch sea salt

Wash and drain millet.

Dry-toast sunflower seeds in a skillet over medium heat until they smell nutty, approximately four minutes.

Bring water and salt to boil. Add millet and seeds. Cover and simmer for 30 minutes. When done, fluff and let sit for 10 minutes. The extra time that grains spend covered makes them nice and soft without getting sticky and smushy. Mix them when they're off the heat.

Serve and devour.

If millet is too dry for you, add more water when cooking. Or add a tablespoon of olive oil when it is done.

To dress up any millet dish, use scallion, parsley, cilantro, or sautéed onion.

Simplicity is beautiful. Now you have proven it!

Millet-Carrot-Hiziki-Burdock: The Reliable Ninja's Favorite Meal

- 2 stalks scallions
- 2 carrots, shredded
- 3-inch piece burdock root
- 1 cup millet
- ¼ inch hiziki, soaked, rinsed, and cut in small pieces
- 6 cups water
- 1 TB. olive oil
- Gomasio to garnish (see recipe for gomasio in "Unbelievable Salad Dressings" section)

Sauté scallions.

Add carrots and sauté for four minutes.

Add chopped burdock.

Add millet.

Add hiziki. Let these aromas get friendly with each other by sautéing them over medium heat for about three more minutes.

Add water and bring to a boil. Cover and cook over low heat for 30 minutes. When done, the millet and hiziki will be moist and full of aroma. Mix and let sit covered for five minutes.

Add olive oil, sprinkle with gomasio, and serve. (Recipe for gomasio is in the "Unbelievable Salad Dressings" section.)

Winter Squash and Millet

1 small onion, chopped
1 small acorn squash, peeled and cubed
3 cups millet
8-inch piece kombu
7½ cups water
1 TB. olive oil

Sauté onion until golden brown.

Add squash; sauté together for three minutes.

Add millet and kombu.

Add water and bring to boil. Cover and simmer for 45 minutes. When done, add olive oil and mix together.

Green Millet

2 cups millet
½ cup parsley
½ cup dill
½ cup scallion
½ bunch collard greens
6 cups water
Sea salt
1 tsp. toasted sesame oil

Wash and drain millet.

Add all green ingredients to water and blend in a blender. If that is too much fuss for you, skip blender and add all the greens into the water with a pinch of salt. Bring to a boil, then cover and cook for 30 minutes.

When done, add a little buzz by adding a heaping teaspoon toasted sesame oil.

If you don't use blender to make green water for cooking, make sure to chop all ingredients finely. It feels as if you are eating a garden!

Optional: To add some style to this, sprinkle with some black pepper or a little cayenne and notice what a difference!

Millet Smashed Potatoes

2 cups millet

7¼ cups water

½ tsp. sea salt

1 small cauliflower, broken into florets

1 sprig tarragon

Lightly toast millet until it smells nutty.

Add water, sea salt, and cauliflower florets.

Bring to a boil. Cover and cook for 30 minutes.

Purée mixture in a blender, adding ¼ cup water.

Garnish with fresh tarragon and serve.

Optional: If you feel a hankering for garlic, add two to three cloves to the blending process. Be bold! Do not apologize!

Millet with Quinoa

1 cup millet

1 cup quinoa

4 cups water

Pinch sea salt

Wash millet. Wash quinoa. If you like the nutty aroma, dry-toast them before adding water to boil.

Add salt and cover.

Cook on low heat for 30 minutes. When done, pull aside and steam for 10 more minutes.

Fluff and serve.

Basic Quinoa

1 cup quinoa
2 cups water
Pinch sea salt

Dry-roast quinoa in a skillet. Stir until it smells nutty.

Add water and salt.

Bring to boil. Cover and simmer for 20 minutes.

Fluff, mix, and serve. Another simple beauty!

Quinoa Poppy

2 cups quinoa
3½ cups water
½ cup poppy seeds
Pinch sea salt

Wash and drain quinoa.

Add water and poppy seeds.

Add salt.

Bring to boil. Cover and simmer for 25 minutes.

Fluff with fork and serve.

Spring Out Quinoa

2 cups quinoa
3½ cups water
1 bag peppermint tea
1 TB. olive oil
Fresh mint, basil, and cilantro

Wash quinoa. Place in water and add peppermint tea bag. Bring to a boil. Cover and simmer for 25 minutes.

When done, add olive oil, fluff, and serve.

Garnish with chopped fresh herbs.

Toasted Pumpkin Seeds with Quinoa

2 cups quinoa

1 cup pumpkin seeds, toasted

4 cups water

1 TB. olive oil

Dry-toast quinoa. Set aside.

Toast pumpkin seeds over medium heat in a skillet. Combine grains and seeds. Place them in water and bring to a boil. Cover and cook for 20 minutes.

When done, fluff, add olive oil, and call it a day.

Quinoa Salad

2 cups cooked quinoa

½ cup radishes, chopped

½ cup cucumber

½ cup celery

½ cup onion, preferably yellow, chopped

½ cup parsley, chopped

½ cup red bell pepper, chopped

1 TB. olive oil

Garnish: cherry tomatoes, shredded garlic cloves

Combine all ingredients through the bell pepper together in a big ball.

Mix well. Top with oil, garnish, and let chill.

Zen Buckwheat

1 cup raw buckwheat

2 cups water

Tiny bit sea salt

Dry-roast buckwheat in a skillet until it turns light brown. Don't wash it or it will stick.

Slowly add buckwheat and salt to boiling water, stirring gently. Cover and simmer for 20 minutes.

For a unique delicacy add two to three umeboshi plums to your cooking, or presoaked shiitake mushrooms.

Buckwheat with Carrot and Arame

½ cup arame
1 large carrot
1 cup raw buckwheat
4 cups water

Soak and rinse arame.

Shred carrot.

Dry-toast buckwheat until nutty and golden brown.

Bring water to boil. Add all ingredients, cover, and simmer for 20 minutes. It is so easy to make and so delicious.

Optional: Add toasted sesame oil, sprinkle with fresh scallion, mix, and shoot for the stars.

Very Russian Buckwheat

1 cup buckwheat
1 cup water
1 cup sauerkraut juice
½ cup sauerkraut

Dry-toast buckwheat.

Mix water with sauerkraut juice and bring to boil. Add buckwheat slowly, then cover.

Simmer for 20 minutes. When done, add sauerkraut. Mix and serve.

To make this dish really awesome, add sautéed onions, minced scallions, garlic, pepper, and, believe it or not, scrambled eggs! Then change your name to Sasha.

Cream of Kasha

1 cup buckwheat
2 cups water
Pinch sea salt
Gomasio
Sunflower seeds

Grind kasha, a nickname for buckwheat, in a blender.

In a saucepan, mix water and kasha, stirring occasionally for smoothness, over medium heat. When it becomes nice and thick, cover and simmer for 15 minutes. When done, sprinkle with gomasio and sunflower seeds.

Soba Noodles Alone

8 oz. soba noodles (100 percent buckwheat noodle)
6 cups water

To make a good soba noodle and avoid overcooking it, here is what you need to know.

Bring noodles to a boil. Do not cover the pot.

Cook for eight minutes.

Rinse with cold water when finished cooking to stop the cooking and seal your noodles.

Add a little olive, sesame, or toasted sesame oil.

Cold Soba Salad

8 oz. Soba, prepared according to basic method
1 bunch sprouted sunflower seeds or watercress
½ cup radishes, chopped
½ cup celery, chopped
½ cup cucumber, chopped

Mix all vegetables and noodles.

For a dressing:

½ cup fresh basil

1 TB. toasted sesame oil

¼ cup tahini

2 TB. tamari soy sauce

2 inches ginger, grated

Juice of ½ lemon

Mix all these ingredients and pour over noodles.

Morning Glory with Oats

2 cups rolled oats

4 cups water

½ cup raisins

¼ cup almonds

¼ cup sunflower seeds

½ tsp. cinnamon

1 grated apple

Grated orange peel

1 TB. raw honey

Combine all ingredients in a pot except the apple, orange peel, and honey. Bring to boil. Lower heat to low.

Shred peeled apple and shredded orange peel into the pot. Continue cooking until water is absorbed and oats become nice and creamy. It will take 15 minutes total.

When it is done, pull aside and stir in 1 heaping tablespoon raw honey.

You can also add more cinnamon or 1 teaspoon vanilla extract.

Muesli

1 cup rolled oats

2 cups almond or soymilk

½ cup dates

½ cup sunflower seeds

Soak all ingredients overnight, and you'll have a delicious breakfast cereal in the morning. Without cooking! It's really, really good!

For extras, add shredded coconut, raisins, or even 1 tablespoon brown rice syrup.

Top Ten Secrets to Cooking the Best Grains

1. Use organic, unrefined whole grains. Quality matters.
2. Always store your grains in an airtight container. To avoid grain creatures, keep peppermint tea bags or bay leaves with the grains.

   Before cooking:
3. Gently wash your grains in cold water; this reawakens dormant energy.
4. Soak them anywhere between 6 to 12 hours. This will activate enzymes, which help with digestion.
5. Discard soaking water.
6. Use a pinch of sea salt when cooking, or add pieces of sea vegetables. This adds a heavenly flavor and nutritional enchantment!
7. One cup grain serves four people.
8. Never touch your grains during cooking. That makes them mushy and they lose water. When cooking is finished, mix them in the pot gently, allowing them to steam covered for 10 minutes.
9. After grains hit the boiling time, reduce heat to low, cover, and let simmer for the suggested time.
10. Remember: Your energy as a cook is just as important as the quality of the food you are using. Therefore, keep your energy positive and joyful around food. What you put into the food will come back when you eat it. Be mindful of that!

Basic Cooking Methods for Preparing Vegetables

Steam

Steaming is a simple way to cook your vegetables. It's also a wonderful way to cut down on excess salt and oil in your diet.

It takes 5 to 10 minutes to prepare green leafy vegetables and 10 to 25 minutes to prepare roots.

What you need is a steaming basket and a pot with lid. Fill the pot with about two inches of water, place the vegetables in the steaming basket, cover, and you're steaming!

Stir-Fry

This is another quick and nutritious way to prepare vegetables. You can stir-fry in oil or in water. This method makes vegetables really tasty because the hot oil seals in flavor. You can use all vegetables, and be aware that the softer the vegetables are, the less time they take to cook.

What you need is a skillet with lid.

If you choose to use oil, heat a skillet and lightly coat the bottom with oil. Add vegetables and sprinkle with a pinch of sea salt; that will make them taste sweeter and draw enough moisture to avoid sticking. If you stir-fry with oil, you can sprinkle water over your vegetables to gain extra steam and heat—and it looks very professional!

If you choose to sautée with water, add one inch water to your skillet and bring to boil. Add thinly sliced vegetables, cover, and simmer for 5 to 10 minutes.

Bake

Baking is great for bringing out the natural sweetness of root vegetables and squashes. All you need is a baking pan, a lid, your vegetables, and an oven, set between 375 and 450°F. Bake for 50 to 60 minutes, depending on the vegetable and its size and thickness.

Quick-Boil

When quick-boiling vegetables, just drop them into boiling water and leave for three to five minutes. This method brings out their flavors, makes them more digestible, and brightens their color. When you're done boiling, rinse with cold water to stop additional cooking and preserve the bright color of the vegetables.

Vegetables You Should Try

Green Leafy

Collard greens	Chards—Swiss chard, red chard, and rainbow chard
Kale—dinosaur kale, purple kale, and lucinato kale	Watercress

continues

continued

Vegetables You Should Try	
Dandelion greens	Parsley
Mustard greens	Lettuces
Beet greens	
Roots and Squashes	
Carrot	Celery root
Parsnip	Acorn squash
Turnip	Butternut squash
Rutabaga	Kabocha squash
Burdock	Burdock root
The Fat Burners	
Daikon radish	Turnip
Leek	Onions
Scallion	Celery
Cabbage Family	
Broccoli	All cabbages
Cauliflower	Brussels sprout
Spices That Are Good to Have	
Bay leaf	Cumin
Oregano	Caraway
Thyme	Mustard seed
Sage	Marjoram
Rosemary	Garlic
Dill	Basil
Fennel	

Steamed Kale

1 bunch kale

Wash and cut kale. Cut into long strips or little bite-size pieces. The stems can also be used.

Place kale in the steaming basket. Put about two inches water into your pot. Place basket into pot. Bring to boil, cover, lower heat, and let steam for five minutes.

Try this with collard greens, bok choy, and mustard greens.

This way, you will know the basic taste of these greens.

How to Make Plain Steamed Vegetables More Exciting

These can be applied to all vegetables you wish to use for steaming

- Add 1 tablespoon olive oil or toasted sesame oil to every 2 cups greens.
- Add 2 bay leaves or 1 teaspoon cumin seeds to the water.
- Sprinkle greens with toasted pumpkin, sesame, flax, or sunflower seeds; almonds; or walnuts.
- Sprinkle greens with fresh herbs: mint, dill, basil, parsley, cilantro, scallions.
- Use tamari soy sauce or umeboshi vinegar to add extra flavor.
- Squeeze fresh lemon juice over them.
- After steaming, quickly stir-fry with a pinch of sea salt, olive oil, and garlic.

Sautéed Greens in Olive Oil

½ bunch mustard greens
½ bunch kale
½ bunch dandelion greens
1 TB. olive oil
2 cloves garlic
Lemon juice to taste

Wash and chop greens.

Heat olive oil.

Add garlic and sauté for a few seconds.

Add greens and stir until all leaves are wilted, about five minutes.

Sprinkle with lemon juice or umeboshi vinegar

Bok Choy Stir-Fry

1 bunch bok choy

½ red bell pepper, chopped

2 cloves garlic

2 TB. olive oil

Pinch sea salt

Wash bok choy and separate greens from stems.

Dice red pepper and garlic.

Heat oil.

Add garlic and sauté for a few seconds.

Add bell pepper. Make sure you stir to prevent garlic from burning.

Add bok choy greens and sea salt first, then add the stems, and stir until they become wilted.

Optional: You can add tamari soy sauce or fresh ginger juice to this dish.

Garlic Steamed String Beans

1 lb. string beans

Water

4 TB. tahini

2 TB. tamari soy sauce

2 cloves garlic

Juice of ½ lemon

Wash and cut beans by chopping both ends off and cutting them in half. Place them in a steaming basket. Add about two inches water to your pot. Bring to boil, cover, and steam for 10 minutes.

Meanwhile, mix tahini, tamari, and shredded garlic cloves into mixture.

When beans are done, put them in serving bowl and pour mixture over it.

Squeeze lemon on top, and you are done.

Steamed Sunny Day

- 2 carrots
- 2 parsnips
- ½ cup sunflower seeds, toasted
- 1 TB. olive oil

Wash and chop carrots and parsnips. The smaller you chop them, the shorter they need to be cooked. Put them into the steaming basket. Add water to the pot and place basket inside. Allow water to boil. Simmer covered on low heat for 5 to 10 minutes, depending on the size of your chunks.

Meanwhile, toast your sunflower seeds in a skillet over medium heat for about four minutes. Pour them into a little bowl.

Add olive oil to your skillet.

Add steamed carrots and parsnips and stir-fry quickly over medium heat for about three to five minutes.

When done, sprinkle with seeds and enjoy.

Dandelion Stir-Fry

- 4 cups dandelion greens
- 1 TB. olive oil or toasted sesame oil
- Juice of ½ lemon

Wash and cut greens.

Heat skillet and add oil.

Add dandelion greens. Keep your flame medium-high, and move your greens so the leaves will be nice and wilted, about five minutes.

When it is done, squeeze lemon over it and serve.

Your liver will love you for this.

Optional: Garnish with gomasio if you wish.

Garlic Gingered Broccoli with Toasted Pumpkin Seeds

3 cloves garlic
5-inch piece ginger root
1 bunch broccoli
6 cups water
1 TB. olive oil
2 TB. tamari soy sauce

Finely grate garlic.

Squeeze the juice from the ginger and set aside.

Wash and cut your broccoli into florets. You can use the stems, but it will take longer to cook.

Add 6 cups water to a pot. Bring to boil. Drop your broccoli in, and quick-boil for about three minutes. Remove from water and give them a quick rinse.

Heat skillet with oil, add garlic and sauté for a few seconds before adding broccoli. Sauté them together, add tamari soy sauce and ginger juice. Let these aromas fuse for about five minutes. That is all it takes to create this beauty!

Optional: Try this dish with cauliflower or Brussels sprouts.

Optional: Garnish with finely chopped fresh tarragon or basil.

Steamed Daikon with Black Sesame Seeds

¼ cup black sesame seeds
2 large daikon radishes
1 TB. olive oil
1 TB. umeboshi vinegar

Toast sesame seeds by placing them on a heated skillet. Stir until they smell nutty. Remove from heat and set aside.

Wash radishes. If you have a little brush you can use for this purpose, you don't have to worry about peeling. Just scrub them well under running water. Cut radishes in half-moon shapes, lengthwise first, then across.

Put them into steaming basket, bring to boil, cover and steam for three minutes.

When done, sprinkle with the toasted sesame seeds, oil, and vinegar.

Sautéed Carrot, Onion, and Arame

- 1 oz. dried arame
- 1 TB. sesame oil
- 1 medium onion
- 2 large carrots
- Water
- 3 TB. tamari soy sauce
- Juice of 1 lemon

Wash and drain arame.

Heat frying pan. Add oil and sauté onions and carrots for three minutes over medium heat.

Place arame over onions and carrots.

Add water *only* to cover the carrots and onions. Bring to boil. Lower the heat and add tamari soy sauce. Cover and simmer for 15 minutes. Add lemon juice. Mix and stir until water evaporates.

Optional: Gomasio and scallions garnish this dish like you wouldn't believe!

Sautéed Cabbage

- 2 cups cabbage
- 1 medium onion
- 1 green apple
- 2 TB. olive oil
- 2 TB. umeboshi vinegar
- 1 tsp. mustard seeds
- 1 tsp. caraway seeds

Finely slice the cabbage, onion, and apple separately.

Heat oil in skillet.

Add and sauté onion.

Add umeboshi vinegar.

Add mustard seeds.

Add caraway seeds.

Sauté together for two minutes.

Add cabbage and sauté for another two minutes.

Add apple.

Cover and simmer on low heat until cabbage is wilted and soft, approximately 10 minutes.

Optional: Garnish with gomasio or toasted pumpkinseeds.

Spring Nishimi

- 2 inches kombu, cleaned, soaked, and cut in little pieces
- 1 acorn squash
- 1 larger daikon radish, cut into 1-inch pieces
- 7-inch burdock root, chopped into small pieces
- 1 sliced carrots
- Pinch salt
- 1 medium onion
- 1 TB. tamari soy sauce

Clean kombu and soak in fresh, cool water for 15 minutes. Cut kombu into little pieces. Place on the bottom of a pot and cover with water.

Add vegetables and sprinkle with a pinch of salt.

Cover and cook over medium heat until steam settles in the pot. Lower heat and cook for 15 minutes. You may want to check your water and add more if water evaporates.

When vegetables are soft, uncover, add tamari soy sauce, and cook five more minutes. Your Japanese nishimi dish is finished!

Optional: Try these combinations: daikon, shiitaki, and kombu; turnip, shiitaki, carrot, kombu; onion, cabbage, daikon, kombu.

Beet-Carrot-Parsnip-Fennel Extravaganza

5 small beets
3 big carrots
2 parsnips
1 fennel bulb
2 TB. olive oil
½ tsp. sea salt

Scrub all the vegetables. Chop vegetables into two-inch pieces and fennel bulb finely.

Preheat oven to 425°F.

Mix vegetables with oil and sea salt. Transfer them to a baking dish.

Bake covered for 30 minutes. Uncover and bake for 15 more minutes.

Roasted Rutabagas with Celery Root

1 rutabaga
1 celery root
2 TB. olive oil
½ tsp. sea salt
1 tsp. fresh rosemary

Preheat oven to 400°F.

Wash and scrub vegetables. Cut them in one-inch-thick rounds. Mix with oil, sea salt, and rosemary.

Cover and bake for 30 minutes. Uncover, flip them over, and bake uncovered for 10 more minutes.

Parsley Stir-Fry

1 big bunch parsley
½ TB. olive oil
2 cloves garlic, finely chopped

This is the simplest stir-fry dish you can get, but what an addition it makes to all dishes—grains especially love it!

Wash and chop parsley.

Heat skillet. Add oil, garlic, and parsley and stir for four minutes. That is all it takes.

Excellent over udon noodles, grain dishes, roasted roots, or by itself.

Optional: Sprinkle with gomasio

Baked Rosemary with Caraway Sweet Potato

- 3 medium sweet potatoes
- 2 TB. olive oil
- 4½ cups fresh rosemary, chopped
- ½ TB. caraway seeds

Preheat oven to 400°F.

Scrub sweet potatoes under running water and cut into big chunks.

Sprinkle baking dish with oil, place the sweet potatoes into the dish, and add rosemary and caraway seeds.

Mix all ingredients together. Cover and bake for 50 minutes.

Optional: Sweet potatoes can be also be baked with cinnamon and 2 tablespoons maple syrup.

Carrot Burdock Power Meal

- 1 onion
- 1 large burdock root
- 1 big carrot
- 1 tsp. olive oil
- Pinch sea salt

Wash and chop vegetables into odd shapes.

Heat skillet with oil. Sauté vegetables together over medium-high heat about five minutes, adding salt and a little sprinkled water. Cover and simmer for 10 minutes on low heat.

Optional: Try adding toasted seeds or fresh parsley for variety.

Roasted Kabocha Squash

1 whole kabocha squash

Preheat oven to 450°F.

Scrub your squash. You can eat the skin if it is cleaned and baked. If that is not for you, cut the squash in half, scoop out the seeds, and cut squash in chunks. However, it is easier to bake the squash with the skin.

Place whole squash or chunks in the oiled baking dish and bake, covered, for 45 minutes. Uncover and bake for 15 more minutes. There's no fuss, and it makes a great meal.

You can wash seeds that you just scooped out and toast them separately with a little sea salt.

Optional: Kabochas are also great baked with cinnamon and with the company of like-minded roots such as sweet potato, rutabaga, parsnip, and/or beets. If you try this colorful combination, add thyme, rosemary, and sage to capture their full flavors.

Boiled Salad

2 quarts water

1 cup carrots, cut diagonally

1 cup sliced leeks

1 bunch broccoli

Bring 2 quarts water to a boil. Boil each vegetable separately for one minute.

Mix them when they are done with a splash of your favorite dressing or lemon juice.

Optional: Here are some creative variations: Try cauliflower, snow peas, dandelion greens, endive, mustard greens, or watercress.

Sweet Sensation

1 cup onions
1 cup carrots
1 cup squash
1 cup water
Pinch salt
1 cinnamon stick or 1 tsp. cinnamon powder

Cut the vegetables into bite-size chunks. Place vegetables in a pot.

Add water and sprinkle with salt and cinnamon. Bring to boil. Lower heat and simmer covered for 20 minutes until vegetables are soft.

Stir vegetables gently and serve.

Pressed Salad

1 cucumber, sliced thin
½ cup onion, sliced
¼ cup dry dulse
1 bunch scallions
½ leek
1 bunch thinly sliced radishes
1 TB. brown rice vinegar
1 TB. umeboshi vinegar

Thinly slice cucumber and onion.

Rinse dulse and slice finely.

Add scallions and only the green part of the leek to the other ingredients.

Mix all ingredients except vinegars. Place them in a bowl or in your baking dish. Press the ingredients to rid the water that is in the vegetables by using natural salt and extra weight.

Cover mixed salad, put a heavy object on top, and let it press for 30 to 45 minutes.

When it is done, squeeze out excess water, add vinegars, and it's ready to serve.

Little Mermaid Salad

2 TB. toasted sesame oil

1 onion, sliced thinly

1 cup arame, washed and drained

Dressing:

1 tsp. rice vinegar

1 tsp. tamari

1 tsp. toasted sesame oil

1 tsp. brown rice syrup

1 tsp. grated fresh ginger juice

½ bell pepper, finely chopped

Heat oil in a skillet. Sauté onion for about three minutes.

Add arame over onion with ¼ cup water and simmer for five minutes.

Remove from heat, add your favorite salad dressing, or mix up the recipe included here and be proud!

Blanched Edamame

2 quarts water

1 (16-oz.) bag frozen shelled edamame

Pinch salt or 8 inches kombu

Bring 2 quarts water to boil. Add salt or 8 inches kombu.

Add 2 cups edamame. Let beans quick-boil for five to eight minutes.

Drain water, sprinkle with salt, and enjoy!

Unbelievable Salad Dressings

A good salad dressing should be healthful and enhance the flavor of your vegetables. You will find it easy to avoid commercial dressings when you discover how simple it is to make delicious, healthy, homemade dressings.

Simple Omega

Mix the following ingredients in a covered jar by shaking:

1 TB. flax seed oil

1 tsp. lemon juice or
1 tsp. umeboshi vinegar

1 TB. balsamic vinegar

1 clove garlic, minced

Italian Dressing

Stir thoroughly and serve:

1 tsp. mustard

¼ cup parsley

3 TB. olive oil

Element Dressing

Stir thoroughly and serve:

1 cup toasted sesame seeds

1 cup water

1 TB. tamari soy sauce

1 TB. umeboshi vinegar

1 TB. brown rice syrup

1 TB. toasted sesame oil

1 TB. ginger, freshly squeezed

Faraway Oriental Dressing

Mix the following ingredients in a covered jar by shaking:

¼ cup tamari soy sauce

¼ cup lemon juice

¼ cup sesame oil

Optional: ¼ cup dulse, chopped

Tahini Lemon Dressing

Stir thoroughly and serve:

2 TB. tahini

1 lemon, juiced

½ cup water

¼ TB. tamari soy sauce

Tofu Mayonnaise

Combine all ingredients in a blender:

18 oz. silken tofu

¼ cup water

2 TB. olive oil

2 TB. brown rice vinegar

¼ cup coriander

1 tsp. sea salt

Great with stir-fry and blanched vegetables.

Parsley Ginger Garlic Dressing

Combine all ingredients in a blender:

1 bunch parsley, chopped

4 inches ginger root

½ cup sesame oil

Diced ginger root

2 cloves garlic

Juice of ½ lemon

¼ cup tamari soy sauce

Gomasio

Combine in blender and sprinkle on everything:

1 cup white sesame seeds

1 tsp. ground sea salt

Tahini Parsley Dressing

Combine the following ingredients in a blender:

½ cup water

2 TB. tahini

1 TB. lemon juice

1 TB. umeboshi vinegar

½ bunch parsley

Pumpkin Seed Dressing

Mix in a bowl. So great on greens!

1 cup pumpkin seed, toasted

2 TB. brown rice vinegar

1 TB. umeboshi vinegar

1 tsp. tamari

1 cup water

Spring Dressing

Combine ingredients in a blender:

½ cup fresh mint, chopped

½ cup fresh parsley, chopped

2 TB. umeboshi vinegar

1 TB. olive oil

2 TB. brown rice vinegar

Beans

Beans are a perfect complement to grains. They are low in calories and fat but high in protein, fiber, the B complex vitamins, and essential minerals, such as calcium, magnesium, iron, phosphorus, potassium, and zinc. Besides being highly nutritious, they are quite easy to prepare.

Basic Bean-Cooking Guide

Wash and clean beans to remove rocks and other unpredictable items.

Soak beans for six hours or overnight. Water should completely cover beans.

Never use soaking water to cook beans with.

Place beans in heavy pot with suggested amount of water. Bring to a boil. Skim the foam off.

Add kombu, bay leaves, garlic, or your choice of herbs.

Cover, lower flame, and simmer for the suggested time.

Only add salt toward the end of cooking, approximately 10 minutes before beans are done.

Pressure-Cooking Guide

Follow steps 1 and 2 from basic cooking guide.

Add legumes to pressure cooker. Also add soaked kombu with legumes.

Cover and bring to pressure. Reduce heat and cook for the suggested time. Remove cover, season, and cook uncovered until water evaporates.

Beans Time Chart

Bean Variety	Minimum Cook Time
Aduki beans	1½ hour
Black beans	45 to 60 minutes
Chickpeas	1½ hours
Kidney beans	1½ hour
French lentils	30 to 45 minutes
Red lentils	20 to 25 minutes
Brown lentils	45 minutes
Split peas	1 hour
Pinto beans	1 hour

Basic Aduki Beans

1 cup beans
5 inches kombu
4 cups water
2 bay leaves
1 tsp. salt

Wash beans.

Soak kombu until it starts to soften up, about 10 minutes. Place kombu and legumes in a pot. Cover with water. Bring to a boil.

Add bay leaves.

Cover and simmer for one hour. After beans have cooked for an hour, add sea salt. Allow beans to cook until they are soft enough for your taste.

To check for softness, take a couple beans from your pot and squeeze them between your thumb and pointer finger. If beans press easily, they are finished. If they feel hard in the middle, they need more time.

Aduki Squash Stew

1 small acorn squash
1½ cups aduki beans
3 inches kombu
3 inches wakame
5 cups water
2 TB. tamari soy sauce

Peel and cube squash into two-inch squares.

Place washed beans and seaweed into pot. Add water. Bring to a boil. Cover and simmer for 30 minutes.

Uncover and add squash cubes. Cover and simmer for 30 more minutes. Uncover, add tamari soy sauce, and stir until water evaporates.

Optional: Try with roots like carrot, parsnip, and turnip. These roots don't need more than 20 minutes to cook with beans.

Red Lentil Burgers

2 cups red lentils
1 medium onion, chopped
2 cloves garlic
½ cup cilantro, chopped
¼ cup hijiki (optional)
3 cups water
¼ TB. dried basil
¼ TB. dried cumin
⅛ TB. turmeric
⅛ TB. thyme
2 TB. tamari soy sauce
1 TB. umeboshi vinegar

Wash and drain lentils.

In a pot, sauté onion, garlic, and cilantro.

Add lentils and hijiki.

Add water, and bring to a boil.

Add spices, cover and simmer for 10 minutes. Uncover, add soy sauce and vinegar, stir, and cover again for another 10 minutes. Stir and simmer vigorously uncovered for three more minutes. It will look mushy because red lentils lose their shape when cooking.

Place them into a bowl to cool until you can form four-inch burgers with them. Sprinkle with scallions and serve with rice and greens.

Minty Green Lentil Watercress Salad

2 cups green lentils
2 bay leaves
4 cups water
¼ cup cumin powder
1 medium carrot, shredded
3 stalks celery, chopped
1 bunch scallions, chopped
1 bunch watercress, chopped
½ cup fresh mint, finely chopped
¼ cup fresh basil, finely chopped
2 TB. olive oil
Pinch sea salt
2 TB. tamari soy sauce
½ tsp. black pepper
2 TB. lemon juice

Wash and drain lentils. Place them in a pot with bay leaves.

Add water and cumin. Bring to a boil. Cover and simmer for 30 to 45 minutes. Check periodically for desired softness. When soft pull aside, but leave covered.

Meanwhile, wash all vegetables.

Chop celery, scallions, and watercress.

Finely chop fresh mint and basil.

In a big bowl, mix chopped vegetables and lentils.

Add olive oil, salt, tamari, black pepper, and lemon juice.

Mung Bean Paté

2 cups mung beans
4 inches kombu
2 cloves garlic
1 medium onion
Pinch sea salt
1 tsp. coriander
1 tsp. cumin
3 TB. tahini butter
2 TB. tamari soy sauce
Juice of 1 lemon

Mung beans are very tasty and delicate beans. Soak them in water overnight, and they'll start sprouting by morning. Start by washing 2 cups mung beans and soak them overnight or for 12 hours. Discard soaking water.

Place beans in a pot with kombu and bring to boil. Cover and simmer for 30 minutes.

Meanwhile, sauté garlic and onion.

Add a pinch sea salt, coriander, and cumin.

When mung beans are finished soaking, transfer them into a blender.

Add the sautéed mixture to it, and blend.

Add tahini and blend more.

Add tamari and lemon juice. Blend for a few seconds. Make sure you turn off blender! And dip your finger into it to taste. Yummy!

Optional: Garnish with chopped parsley or scallions.

Optional: You can make this paté without cooking the mung beans and soaking them for two days instead. If you do so, change the water two to three times during the soaking period. Wonderful for summer!

Black Bean Salad

2 cups black beans
5 inches kombu
3 cups water
1 tsp. coriander
2 tsp. cumin
1 onion, diced
1 red, green, and yellow bell pepper
2 cloves garlic
1 TB. olive oil
1 TB. cilantro, chopped
1 tsp. sea salt
Pinch cayenne
Juice of 1 lemon

Wash beans. Place them in a pot with kombu. Add water and bring to boil. Cover, add spices and simmer for an hour.

Chop onions and peppers, and mince garlic.

Meanwhile, sauté onions, garlic, and cayenne.

Mix beans with sautéed onions.

Add peppers, olive oil, and salt.

Garnish with cilantro and squeezed lemon juice.

Basic Chickpeas

1 cup chickpeas
5 inches kombu
2 cups water

Wash chickpeas. Place them in pressure cooker with kombu and cover with water. Bring to a boil. When steam settles, lower heat and cook for one hour.

Optional: You can make delicious salads by adding chopped vegetables, sea vegetables (hijiki, arame), onions, scallions, dill, basil, and other herbs.

Or you can make:

Hummus

2 cups cooked chickpeas
⅓ cup chickpea water left over from pressure cooker
3 TB. tahini
3 cloves garlic to taste
½ tsp. sea salt
2 TB. lemon juice
¼ tsp. black pepper
⅛ tsp. cumin
⅛ tsp. coriander

Place all ingredients in a blender, blend, and celebrate!

Optional: To add a rich flavor, sauté onions and garlic and add to blender.

Optional: Garnish with dill, scallions, or cilantro.

French Lentil Shiitake Salad

8 to 10 shiitake mushrooms
2 cups French lentils
4 cups water
8 inches wakame or kombu
1 tsp. thyme powder
1 tsp. dried rosemary
2 medium parsnips
2 TB. tamari soy sauce
4 cloves garlic, diced
1 bunch scallions
½ cup basil
3 TB. olive oil
½ tsp. black pepper

Soak the mushrooms for at least 20 minutes and cut up.

Wash lentils. Put in pot with water.

Add kombu or wakame to water and bring to a boil.

Add thyme powder and rosemary to lentils.

Cover and simmer for 15 minutes on low heat. Uncover and add chopped parsnips and mushrooms. Cover and simmer for 15 more minutes.

Transfer lentils into a big bowl; add soy sauce, garlic, scallions, basil, olive oil, and black pepper.

Optional: You can add leftover or fresh grains of any kind to this wonder.

Soups

There's nothing like homemade soup on a cold, winter day to warm us up and deeply nourish us. The ingredients in a well-prepared soup blend together and become more digestible and delicious than when served alone. Start cooking your soup before you prepare the other parts of your meal. Give the soup lots of cooking time and love. It will be well worth it!

Easy Breezy Soup

Try any of these vegetable combinations to create a simply delicious soup:

Carrot, parsnip, celery, and parsley

Squash, carrot, and ginger-beet

Broccoli, onion, and leek

Daikon, leek, parsley, and carrot

Mustard greens, leek, and onion

Kale, cabbage, and parsnip

Use 5 cups water. Bring to boil.

Add 1 cup of each vegetable you choose to cook.

Cover and simmer for 20 minutes.

Optional: Garnish with parsley and scallions.

Optional: To add richness to soup, sauté 1 medium onion and add to water before cooking.

Carrot Ginger Soup

6 carrots

Onion

1 tsp. sea salt

4 cups water

6-inch piece fresh ginger, grated

Parsley to garnish

Wash, peel, and cut carrots and onion into chunks. Place vegetables and salt in a pot. Add water.

Bring to a boil. Cover with a lid. Simmer on low heat for 25 minutes.

Transfer soup into blender, adding water to achieve desired consistency. When blending is done, squeeze juice from grated ginger and add to soup. Garnish with parsley.

Optional: Sauté vegetables before cooking for extra flavor.

Optional: Substitute carrots with squash, parsnips, or beets. Squash and beets need 35 to 40 minutes to cook.

Green Lentil Soup

1 cup green lentils
1 TB. olive oil
1 medium red onion
1 carrot
1 parsnip
5 inches burdock (optional)
2 bay leaves
4 cups water
Pinch sea salt
1 TB. cilantro, chopped
2 TB. tamari soy sauce

Wash lentils. Bring to a boil and cook for a half-hour. Put aside.

In another pot, add oil and sautéed onions.

Add carrot, parsnip, burdock, and bay leaves. Sauté together.

Pour lentils over sautéed mixture; add water and a pinch of sea salt.

Bring to boil and cover. Turn heat to low, and simmer for 45 minutes.

Add chopped cilantro and tamari to finished dish.

Mix and serve.

Hot Borscht with Tofu Sour Cream

5 large beets
2 purple onions
2 TB. sea salt
10 cups water
5 inches wakame
2 TB. lemon juice
½ cup maple syrup
½ cup apple cider juice

Wash beets. Peel and cut into two-inch chunks.

Sauté onions with salt.

Add beets to onions and sauté them together. This will bring out extra sweetness from beets.

Add water, wakame, and lemon juice, and bring to a boil. Cover and cook for 20 minutes on low heat.

Add maple syrup and cider, and cook for 20 more minutes.

Serve with tofu sour cream.

Tofu Sour Cream

1 lb. tofu
½ cup scallions, chopped
1 TB. umeboshi vinegar
1 TB. lemon juice
2 TB. olive oil

Place all ingredients in a blender except olive oil. Blend until smooth. When cream reaches desired smoothness, pour olive oil *slowly* into cream.

Store in refrigerator for up to three days.

Optional: Garnish with dill.

Split Pea Soup

- 2 cups split peas
- 8 cups water
- 6 inches kombu
- 1 large onion
- 2 large carrots
- 2 parsnips
- 2 TB. tamari soy sauce
- ½ cup fresh dill

Soak peas for several hours, then wash.

Add water and kombu.

Bring to boil, and skim off any foam.

Add onion and more water, if necessary, and simmer on low heat, for 30 minutes.

Add all other vegetables, tamari, and dill and simmer, covered, for an additional 30 minutes.

Creamy Chestnut Squash Soup

- 2 large kabocha squashes
- 2 medium onions, chopped
- 1 cup dried chestnuts
- 6 cups water
- 2 tsp. sea salt

Peel and cut squash into 2-inch chunks.

Chop onions.

Soak chestnuts for 20 minutes.

Preheat oven to 400°F.

When chestnuts are soft, cut them in half. Put them in a baking dish, cover, and bake for 30 minutes.

Sauté onions in a pot for four minutes until they are nice golden brown.

Add kabocha chunks.

Sauté together for a few minutes.

Add water and salt. Bring to a boil, and cover. Simmer for 30 to 45 minutes.

When chestnuts and squashes are ready, transfer them into a blender and, using the squash cooking water, blend to desired consistency.

Mighty Miso Soup

8 inches wakame

5 inches kombu

1 large onion

1 medium daikon radish

5 cups water

1 to 2 TB. barley miso

1 cup scallions or leeks, chopped

Wash and soak wakame and kombu for three minutes until soften, and cut into little pieces.

Chop onion and daikon.

Add all vegetables and sea vegetables to water and bring to boil. Do not add miso yet! Reduce heat, cover, and simmer soup broth 10 more minutes.

Meanwhile, remove 1 cup liquid from the pot, dissolve miso paste, and return to the pot.

Mix in miso and reduce heat to very low and cook for two to three minutes. Do not boil.

Optional: Garnish soup with scallions or leeks.

Shiitake Miso Soup

8 inches kombu

6 cups water

8 dried shiitake mushrooms

1 cup fresh shiitake mushrooms

1 medium onion

1 large carrot

6 to 8 TB. unpasteurized barley miso

Parsley

Soak kombu in water.

Soak mushrooms in water for 20 minutes. Remove stems and thinly slice all mushrooms.

Dice onion and slice carrot into rounds.

Place water into a soup pot, and bring to boil.

Add onions, carrots, shiitake mushrooms, and kombu. Reduce heat to low, cover, and simmer for 10 minutes.

Remove 1 cup liquid from the pot, dissolve miso paste, and return it to pot. Turn heat very low and cook for two to three minutes.

Garnish with parsley and serve.

Optional: You can add other vegetables to your miso soup, including daikon radish, leek, snow peas, mustard greens, and Chinese cabbage.

Creamy Salmon Soup with Mustard Greens

1 large onion

1 to 2 TB. olive oil

1 small daikon radish

1 carrot

½ bunch mustard greens

10 to 12 oz. baked salmon

3 TB. lemon to taste

1 tsp. fresh dill, chopped

Sauté onion with olive oil.

Add daikon and carrot, and sauté for a few minutes.

Add dill and keep sautéing.

Add mustard greens and sauté another minute.

Add water, bring to boil, and cover. Simmer for 15 minutes.

Transfer soup to the blender, add baked salmon, and blend.

Return to pot and cook 10 more minutes.

Squeeze lemon juice into the soup before serving.

Sprinkle with fresh dill.

Optional: For a vegan variation, use 8 ounces tofu instead of salmon.

Creamy Broccoli Soup

- 1 bunch broccoli
- 5 cups water
- 1 small onion
- 2 cloves garlic
- 1 cup cooked brown rice
- 2 TB. barley miso

Wash broccoli and separate stems from florets.

In a pot, bring water to a boil.

Add broccoli stems and onion.

Mince the garlic and add to the pot. Reduce heat and simmer for 10 minutes.

Take 2 cups liquid out of the soup, and add brown rice and miso. Put liquid in the blender and blend. When it's smooth, return to the pot.

Add broccoli florets and cook 10 more minutes.

Tofu, Tempeh, and Natto

Tofu, tempeh, and natto are soy products. Tofu is mashed and curded soybeans that are formed into a soft, rectangular-shape block. It is rich in protein and phytochemicals. Tempeh is whole fermented soybeans packed into a patty. Like tofu, it is rich in phytochemicals, but tempeh is also abundant in friendly bacteria and soluble fibers. Natto is a delicious soybean condiment for grains that can be purchased in most natural foods stores.

Marinated Tofu Stir-Fry

8 oz. tofu

Marinade:

- 1 TB. ginger juice
- ½ TB. tamari soy sauce
- ½ TB. brown rice vinegar
- ½ TB. toasted sesame oil
- ½ cup fresh cilantro
- 2 cloves shredded garlic

Drain excess liquid from tofu by placing a heavy object over it. Cut tofu into 1-inch squares after draining. Set aside and prepare marinade by mixing all ingredients except oil.

Marinate tofu for 30 minutes or overnight.

Heat oil in skillet.

Add tofu and quick stir-fry until tofu becomes golden brown.

Scrambled Tofu

1 block tofu, squeezed and crumbled

2 TB. olive oil

⅛ tsp. paprika

½ tsp. tamari soy sauce

⅛ tsp. turmeric

1 red onion

½ red bell pepper

1 TB. umeboshi vinegar

Dash black pepper

Press tofu and crumble into small pieces.

Heat oil.

Add paprika, tamari, and turmeric. Sauté for a few minutes.

Chop onion and bell pepper.

Add vegetables, umeboshi vinegar, and black pepper.

Cook for five minutes until mixture thoroughly heats and flavors blend.

Optional: Use alfalfa sprouts or fresh parsley to garnish.

Sunrise Tofu

4 sliced carrots

6 beets, cut in chunks

1 large onion

Water

1 lb. tofu

½ TB. salt

In a large pot, add carrots, beets, and onion. Add 1 inch water.

Cook covered over medium-low heat for 20 minutes.

Add tofu and cook 10 minutes more.

Season with salt.

Optional: For variation, try adding ½ cup wakame when cooking vegetables.

Marinated Baked Tofu

1 lb. firm tofu

Marinade:

- 1 small onion
- 3 cloves garlic
- 1 cup water
- 4 TB. olive oil
- 2 TB. brown rice vinegar
- 1 TB. dried basil
- 1 TB. dried oregano
- ½ tsp. black pepper
- Dash cayenne pepper or paprika

Rinse and press tofu. Cut 1-inch-square chunks.

Prepare marinade by mixing all ingredients.

Add tofu and marinate in the refrigerator for one hour.

Preheat oven to 375°F. Place marinated tofu on baking sheet and bake until golden and crispy, about 10 to 15 minutes.

Garnish with parsley or sprouts.

Easy Tofu Dip

- 1 block tofu
- 1 tsp. sea salt
- ¼ cup water
- ½ TB. grated ginger juice
- ½ bunch scallions
- 1 TB. toasted black sesame seeds
- 1 TB. sesame oil

Put all ingredients into a blender except sesame oil. Blend mixture. When smooth, add oil. Blend more.

Serve with raw cut vegetables or over rice.

Tofu Mayonnaise

- 1 block tofu
- 1½ TB. apple cider
- 1 TB. natural stone-ground mustard
- 1 tsp. brown rice syrup
- ⅛ tsp. turmeric

Blend all ingredients together.

Refrigerate in a covered glass jar up to four days.

Great with baked fish or baked root vegetables.

Tofu Whipped Cream

- 1 cup silken tofu
- 1 cup raw honey or ½ cup brown rice syrup
- 1 tsp. vanilla extract
- 1 TB. safflower oil

Drain tofu and pat dry.

Place all ingredients in a blender and blend until smooth.

Refrigerate.

Awesome with baked pears, apples, banana, or squash.

Strawberry Tofu Pudding

- 3 to 4 dates
- ½ cup silken tofu
- 1 cup strawberries
- 1 TB. shredded coconut
- 1½ TB. raisins

Pit and soak dates.

Blend tofu, dates, and strawberries.

Refrigerate for 1 hour.

Mix coconut and raisins and sprinkle over tofu, date, strawberry mixture.

Banana Mint Tofu Pudding

- 2 to 3 dates
- 2 ripe bananas
- 1 cup silken tofu
- ⅛ tsp. cinnamon
- ⅛ tsp. nutmeg
- Mint leaves

Pit and soak dates.

Blend all ingredients except mint leaves.

Refrigerate for 1 hour.

Garnish with mint leaves.

Marinated Tempeh

- ½ cup water
- ½ TB. grated ginger
- 1 TB. curry powder
- 1 TB. umeboshi vinegar or ½ tsp. sea salt
- 1 tsp. cumin
- 1 TB. brown rice vinegar or ½ cup lemon juice
- 8 oz. tempeh
- 1 TB. olive oil

Mix ingredients except tempeh and olive oil for marinade.

Cut tempeh into strips.

Soak tempeh for 30 minutes in this sauce.

Heat skillet, add oil, and quickly stir-fry tempeh until golden brown.

Very Versatile Mashed Tempeh

½ cup soaked arame, hiziki, or dulse

2 TB. water

8 oz. tempeh

1 small red onion

½ bunch scallions, finely chopped

½ cup celery or red bell pepper, finely chopped

1 TB. tahini

1 TB. grated ginger juice

1 TB. lemon juice

Rinse, wash, and soak sea vegetables for 20 minutes. However, you don't have to soak dulse.

Place tempeh, onion, scallions, and sea vegetables in a pot with water and cook until they are soft and water has evaporated, about 30 minutes.

Transfer to a bigger bowl and mash with potato masher or fork.

Add celery or pepper, tahini, ginger juice, and lemon juice. Mix very well.

Wrap this in …

- Whole-sprouted grain burritos.
- Whole-wheat pita bread.
- Blanched, dark green vegetable leaves: collard greens or kale.

Basic Natto

1 cup natto

⅛ tsp. sea salt

1 TB. tamari

2 to 4 TB. scallion, chopped

1 clove garlic

Simply mix, serve, and enjoy!

Sea Vegetable Dishes: Kings and Queens from Under the Sea

Arame Mild, semisweet flavor, thin but firm texture. Great as a side dish, but especially loves buckwheat.

Dulse Savory tasting, brownish-green colored stalks. Wonderful for roasting with seeds and as a condiment.

Hiziki Robust in flavor and black in color, a great side dish.

Kombu Light in flavor and chewy. Expands and softens when soaked. Excellent food tenderizer and helps with the digestibility of beans. Adds a sweet flavor to root vegetables. Creates wonderful stocks and soups.

Nori Paper-thin, dark green sheets from pressed sea vegetables. Nori has a similar flavor to tuna. Originally used as sushi wrap. Can use nori flakes as a condiment.

Wakame Delicate long green strips. Has a sweet flavor. When soaked, wakame expands a great deal, so cut it into small pieces. Wakame loves the company of carrots and parsnips and adds a sweet taste to all legumes.

How to Clean Sea Vegetables

Arame and hiziki need to be washed and soaked. Soaking will help with their digestion, cooking time, and taste. Arame requires less time to soak. Some people do not soak them at all. Great with soba noodles!

Put sea vegetables into a bowl filled with cold water.

Move fingers through the stems, like shampooing.

Discard this "first wash" water.

Rinse through again.

Fill the bowl with cold water.

Place sea vegetables in, and let stand for 15 to 20 minutes.

Use the water to add to your houseplants or rinse your hair with and watch it grow!

Basic Cooking for Arame and Hiziki

After washing and soaking, place them in a pot. Add water and bring to a boil. Lower heat and simmer for 5 to 10 minutes.

Arame Buckwheat with Onion

½ cup arame

2 cups water

1 cup dried buckwheat groats

¼ cup onion, chopped

Soak and rinse arame.

Bring water to boil.

Slowly add buckwheat and arame and onion (if you have no time to chop, try just adding the onion whole).

Cover; simmer on low heat for 20 to 25 minutes. Do not stir during cooking.

Eggs with Arame

½ cup arame

2 eggs (free-range organic)

Pinch sea salt or ½ tsp. tamari

Soak and wash arame.

Prepare eggs as you like best.

Cook arame in 1 inch water for 5 minutes or until it moistens and heats nicely through.

Remove arame from water. Add to eggs.

Splash with sea salt or tamari.

Garnish with scallions or parsley.

Collard Leaves Stuffed with Carrot Arame Salad

6 cups water

3 whole large collard leaves

½ cup arame, washed soaked

4 large carrots, shredded

3 zucchini, shredded

2 TB. olive oil or toasted sesame oil

1 bunch scallions, finely chopped

For the dressing:

3 TB. tahini sesame butter	1 tsp. ginger juice
2 TB. lemon juice	¼ tsp. salt (optional)

Boil water. Drop leaves in for one minute. Remove leaves from water, spread them out, and let them cool.

Meanwhile, soak arame for 10 minutes.

Shred carrots and zucchini.

Heat oil in skillet. Sauté scallions for two minutes.

Add carrots and zucchini and sauté for four minutes.

Add arame and sauté for three more minutes.

Mix dressing ingredients, adding a little water to achieve desired consistency.

Mix dressing with vegetables. Toss with toasted sesame oil and a splash of lemon juice.

Separate 1 cup servings of vegetables and place them on the cooling collard leaves. Wrap them up nicely. You have just created something extraordinary!

Note: To professionally seal the wraps, cut four red pepper rings and pull over wraps.

Kombu Stock

2 (12-inch long) pieces kombu	6 cups water

Soak kombu in cool water for 30 minutes. When soft, cut into 1-inch pieces. Discard soaking water and add fresh water.

Cook over medium heat for 20 minutes.

Remove kombu. (Use this water to prepare soup or grain dishes.)

Optional: For variation, add one or two nori sheets, or add 1 cup onion or scallions, or add a 10-inch piece burdock root.

Wakame with Greens

½ cup wakame, soaked and chopped
1 bunch green leafy vegetables (collard, kale, or mustard greens)
1 TB. olive oil
Dash salt
Juice of ½ lemon or brown rice vinegar
2 TB. gomasio

Wash, soak, and chop wakame into small pieces.

Wash and chop greens into bite-size pieces.

Cook wakame in a small amount of water until it becomes tender, about five minutes.

In a skillet heat oil, add greens, and sauté for five to seven minutes until leaves wilt.

Add a dash of salt to sauté.

Add soaked wakame and lemon juice.

Sauté together for three to five more minutes.

Sprinkle with gomasio and serve.

Kung-Fu Salad

1 cup hiziki
1 tsp. sesame oil
2 tsp. umeboshi vinegar
1 yellow or red pepper
4 stalks scallions
1 carrot
½ cup corn grit or snowpeas
5 to 6 cherry tomatoes (optional)
2 cloves garlic
1 TB. ginger juice

Rinse and soak hiziki.

Cook hijiki in water for 10 minutes. Drain from water and allow to cool.

Chop and mix all ingredients.

Add hiziki.

Note: This was Bruce Lee's favorite meal.

Dulse Sandwich with Sesame Spread

2 slices sprouted whole-grain bread
1 handful sprouts
½ cup soaked or toasted dulse
1-inch slice tofu
Handful lettuce leaves
1 to 3 onion rings
1 forkful sauerkraut

Spread:

1 tsp. tahini
1 TB. lemon juice
½ tsp. sesame seeds, toasted

Mix spread ingredients. Spread on bread slices. Arrange the ingredients in a tasteful order. Pack it, and you're ready to go.

Gingered Hiziki with Mung Bean Sprouts

1 cup hiziki
1 TB. tamari
1-inch piece ginger
2 TB. olive oil
2 cups water
1 cup sprouted mung beans

Wash and soak hiziki.

Peel and slice ginger into thin slices.

Heat oil and add hiziki and tamari. Sauté for two to three minutes.

Add ginger and sauté two more minutes.

Add water. Cover and simmer for five minutes.

Add sprouted mung beans. Mix and stir for a few more minutes.

Remove from heat and serve.

Salmon and White Fishes

After-Work Salmon Cake

- 4 oz. cooked salmon
- 6 rice crackers
- ½ onion
- 2 cloves garlic
- 1 TB. lemon juice
- Dash black pepper
- Dash coriander
- 1 TB. olive oil

Break salmon and rice crackers into small pieces.

Mix all ingredients except oil together. Create several small patties.

Refrigerate for one hour.

In a skillet, heat oil on high. Quickly fry both sides of each patty for two minutes.

Serve with brown rice and lemon slices.

Easy Salmon

- 1 carrot, diced
- ½ cup parsley, chopped
- 4 bay leaves
- 6 oz. salmon
- 2 slices lemon
- Dash black pepper

Place diced carrot in a pot.

Add parsley and bay leaves.

Place fish on top, adding lemon and pepper. Add enough water to cover fish. Bring water to a boil. Reduce heat and simmer uncovered for five minutes. Turn heat off and let sit for five more minutes.

Ideal Dill Fish

1 lb. cod fish fillet
Dash black pepper
½ tsp. sea salt
½ cup fresh dill
Water to cover fillets
1 TB. freshly squeezed lemon juice

Rinse fish. Season with pepper and salt.

Add dill.

Heat water in skillet. Drop in fish and cook until it's soft, about five to seven minutes. Add freshly squeezed lemon juice.

Pan-Fried Tilapia

5 tsp. olive oil
1 tsp. fresh sage (optional)
8 oz. tilapia
½ lb. fresh shiitake mushrooms
Salt and pepper to taste
1 tsp. grated lemon peel
3 stalks scallions, sliced small

In a skillet, heat oil and sage, if used, and cook fish two to three minutes per side.

Add shiitake mushrooms.

Season with salt and pepper.

Cook for another minute or two.

Remove from heat and sprinkle with lemon peel. Garnish with scallions.

Serve and enjoy.

Ginger Halibut with Shredded Daikon Radish

- 8 oz. halibut
- 1 TB. tamari soy sauce
- 1 TB. sesame oil
- 1 tsp. fresh ginger juice
- ½ cup water
- 1 TB. olive oil
- 1 cup daikon radish, shredded

Wash fish.

Mix tamari, sesame oil, ginger juice, and water.

Marinate fish in ingredients for 30 minutes.

Heat olive oil in a skillet. Add marinated fish with the leftover marinade. Cook uncovered for about five minutes.

Serve with shredded daikon radish.

Desserts

A good dessert does not take a long time to make.

A good dessert has all natural sweet taste without added sugars and chemicals.

A good dessert can be enjoyed with no guilt and lots of fun.

A good dessert in small amounts is totally satisfying.

A good dessert is a chance to be creative in the kitchen.

A good dessert can act as an invitation for imagination!

In these recipes you will find a magic ingredient called kuzu. You might not know it well, but kuzu is originally from Japan and is very useful in preparing thick sauces, creams, or soups. The white root is made into a powder that dissolves in cold water and becomes thick in hot water. It is all-natural with no bitter or sweet aftertaste. It also has tremendous healing benefits in that it is alkalinizing and soothing, relieves stomach aches, controls diarrhea, is great for colds and flu, and restores overall strength. Please enjoy experimenting with this amazing powder.

Fresh Fruit Salad

1 banana
1 apple
1 pear
1 mango
Fresh chopped mint
1 TB. almonds
1 TB. cashews
1 TB. sunflower seeds

Wash and peel fruit. Cut into little pieces.

Mix together and top with fresh mint, nuts, and seeds.

Optional: Squeeze fresh lemon over salad for a tangy, zesty taste.

Honey Baked Grapefruit with Coconut Glazed Carrots

2 large grapefruits
2 TB. honey
2 large carrots
1 tsp. coconut oil

Cut grapefruit in half. Smear honey between the little sections.

Put the halves in a baking dish and bake at 300°F for 10 minutes.

Wash and shred carrots.

Heat skillet and drop in a teaspoon coconut oil.

Add shredded carrots. Sauté over medium heat for five to seven minutes.

Spoon carrots over grapefruit halves and serve.

Almond Apricot Cups

1 cup dried apricots
¾ to 1 cup water
2 TB. honey
Juice of 2 inches grated ginger, squeezed
Grated lemon peel
½ cup shredded almonds

Soak apricots overnight.

Blend apricots in a blender with ¾ to 1 cup water and honey.

Add ginger and lemon peel.

Transfer into little cups and top with almonds.

Black Bean Chocolate Mouse

- 1 cup black beans, soaked and washed
- 1 TB. cinnamon
- 1 TB. vanilla extract
- ½ cup raisins
- ½ cup grated coconut
- 1 TB. brown rice syrup
- 2 TB. tahini
- ½ cup carob powder
- ½ cup amesake

Soak beans overnight and wash. Cook beans with cinnamon, vanilla extract, raisins, and coconut for one hour.

Transfer to a blender. Add remaining ingredients. Blend until very smooth.

Chill and serve.

Cinnamon Baked Squash

- 1 medium squash
- 1 to 2 TB. honey or maple syrup
- 1 TB. cinnamon

Wash and cut squash in half. Take out seeds. Put squash in a baking dish, add your choice of sweetener, and sprinkle with cinnamon.

Bake for 45 minutes at 450°F.

Easy, simple, and very rewarding!

Note: Oftentimes squashes are sweet enough just by themselves, so you can omit the sweeteners or use apple chunks, raisins, figs, or dates.

Note: You can save the seeds and soak and bake them.

Ball-o-Nuts

- 6 dates, soaked
- ½ cup sesame seeds
- ¾ cup almond
- ½ cup sunflower seeds
- ½ cup apple juice
- ½ cup brown rice syrup
- ½ cup rolled oats, soak them with dates
- ¾ cup poppy seeds

Soak dates.

Add all ingredients except poppy seeds into a blender. Blend until chucks become very small, but are still apparent.

Form little balls. Cover in poppy seeds.

Note: You can also squeeze lemon juice or ginger juice into the paste.

Ginger Squash Cookies

- 1 small acorn or kabocha squash
- Dash cinnamon
- 1 cup whole-wheat flour
- ½ cup amesake
- Dash nutmeg
- 1 tsp. grated ginger
- ½ cup brown rice syrup
- ¼ cup corn oil

Cut squash in half and take out seeds.

Bake with cinnamon at 450°F in a covered baking dish for 35 to 40 minutes. When squash is done it will soften enough to mash.

Mash and mix with the rest of the ingredients.

Form little patties, or use a cookie cutter.

Arrange cookies on a lightly oiled baking sheet, and bake at 350°F for 25 minutes.

Sweet Chestnut Pudding with Tofu Whipped Cream

1 cup dried chestnut, soaked overnight
1 cup chopped dates
Dash cinnamon
1 TB. raw honey or
2 TB. brown rice syrup

For whipped cream:

1 block silken tofu
1 tsp. vanilla extract
½ cup raw honey
1 TB. safflower oil

Soak chestnuts overnight. Drain chestnuts and rinse. Place them in a pot to cook with dates and cinnamon for 45 minute to 1 hour.

While the chestnuts are cooking, prepare the whipped cream as follows:

Drain and rinse tofu.

Put silken tofu, vanilla extract, and honey in a blender. Blend tofu to a creamy consistency.

Slowly pour in the oil, and blend on low speed for 1 more minute.

Remove from blender and refrigerate.

When the chestnuts are finished, transfer ingredients to the blender.

Add sweetener and blend until smooth.

Chill or serve immediately with the whipped cream on top.

Optional: Top with walnuts or orange grinds.

Blueberry Cream

1 cup blueberries
1 cup organic apple cider
2 TB. kuzu
½ cup water
½ TB. maple syrup

Wash berries.

In a saucepan, cook berries with apple cider until they began to soften, about 10 minutes. Pull them aside.

Mix kuzu with water and let it dissolve. Add kuzu to cooked blueberries.

Stir and cook over low heat until mixture thickens, about 5 minutes.

Add maple syrup, stir, and remove from heat.

Chill and serve when ready.

Optional: Top with whole blueberries.

Indian Rice Pudding

- 1 cup basmati rice
- 4 cinnamon sticks
- 1 tsp. cardamom powder
- 1 tsp. whole cloves
- 2 tsp. nutmeg
- Shredded lemon peel
- Shredded orange peel
- 2 TB. raw honey
- 2 cups water
- ½ cup golden raisins
- ½ cup shredded coconut
- Almond milk
- ½ cup pistachios or almonds

Place rice, spices, peels, and honey in water and bring to a boil. Cover and simmer for 20 minutes. When rice is done, transfer it into a bowl and remove cinnamon sticks.

Mix in raisins and shredded coconut.

Pour almond milk slowly into mixture to achieve your desired consistency.

Add pistachios or almonds.

This dessert can use all kinds of imaginative decorations and spices.

Optional: You can be very exotic and add 1 teaspoon rosewater also.

Baked Bananas

- 4 firm bananas
- 1 tsp. olive oil
- 1-inch piece grated ginger
- ½ cup raisins
- 1 TB. cinnamon
- ½ TB. nutmeg

Peel and cut bananas in half lengthwise. Oil a baking pan and arrange bananas.

Sprinkle with spices and raisins, cover, and bake at 375°F for 10 to 15 minutes.

Remove from heat and serve. Wonderful with a bit of chocolate sauce.

Coconut Date Cookies

- 6 to 7 dried pitted dates
- 2 cups whole-wheat flour
- 1 cup rolled oats
- ½ cup olive oil
- ½ cup maple syrup
- ½ TB. cinnamon
- ¼ TB. nutmeg
- ½ cup shredded coconut

Soak dates in 1 cup water for 30 minutes.

Put all ingredients including date soaking water in a blender. Blend into dough.

Form little balls, and smash them into cookies. Place them on a lightly oiled cookie sheet and bake at 375°F for 10 minutes. Turn them over and bake 10 more minutes.

Romantic Apple Sauce

- 1 lb. red delicious apples
- 4 to 5 cinnamon sticks
- 1 TB. lemon peel
- ½ TB. maple syrup
- 1 tsp. whole cloves
- 1 cup organic apple cider
- Dash cinnamon powder

Peel and chop apples. Place apples in a pot with enough water to cover them.

Add cinnamon sticks, lemon peel, maple syrup, and whole cloves. Bring to a boil, cover, and simmer for 15 to 20 minutes.

Place apples in a blender. Blend with enough cider to have mixture become nice and silky.

Add a dash cinnamon or ginger juice when it is ready to serve.

Optional: Combine more than one fruit like pears and oranges to create something very special.

Mochi

Mochi is another widely used food in Japan. It is made from pounded sweet rice, cooked for a long time with a lot of water, until rice becomes really sticky, and well pounded into a flat rice cake to be reheated and consumed. It is very sweet and has great healing advantages. Good for breast-feeding, for anemia, for overcoming general fatigue, and for weight gain. It can be found next to the refrigerated tofu and tempeh.

Basic Mochi Preparation

1 block mochi

Cut mochi into 2-inch cubes. Place cubes on a lightly oiled skillet.

Cover and heat over a low flame until the pieces are puffed and expanded, about 10 minutes. Eat them just as is or with fruit spread, squashes, beans—use your imagination!

Raisin Pudding

1 cup raisins

1½ cup water

1 tsp. cinnamon

2 TB. kuzu

In a saucepan, cook raisins in water for 15 minutes. Add cinnamon if you wish. When finished cooking, blend in blender.

Meanwhile, dissolve kuzu in 1½ cup water and mix in with blended raisins.

Bring mixture back to saucepan and cook over medium heat for five more minutes.

Dash with cinnamon and serve.

Wheat-Free Sunflower Crunchies

1 cup sunflower seeds

½ cup sesame seeds

1 TB. poppy seeds (optional)

1½ TB. olive oil

1 TB. maple syrup

Combine sunflower, sesame, and poppy seeds and blend.

Add olive oil and maple syrup and blend again.

Roll dough into several long fingers. Put them on a lightly oiled baking sheet.

Bake them for 15 to 20 minutes at 375°F.

These cookies are great with raisin sauce.

Poached Pears with Walnuts

4 to 6 pears

½ cup maple syrup

½ cup walnuts

Dash cardamom

Wash and cut pears in half.

Lightly oil a baking dish, and arrange pears.

Splash with maple syrup and sprinkle with walnuts and cardamom.

Bake covered at 350°F for 15 to 20 minutes.

Excellent with tofu whipped cream. (Recipe in "Tofu, Tempeh, and Natto" section.)

appendix A

Your Basic Shopping List

First, spot a health food store around your neighborhood, because many of these items have not yet been introduced to supermarket consumers. Buy organic! It makes a difference!

Grains

Amaranth

Short-grain brown rice

Long-grain brown rice

White basmati rice

Brown basmati rice

Wild rice

Barley

Millet

Rolled oats

Buckwheat—raw, not toasted

Soba noodles

Udon noodles

Quinoa

Beans and Soy Products

Aduki beans

Lentils (brown, red, and French)

Garbanzo beans

Black beans

Split peas

Tofu—fresh soybean curd made of soybean and natural sea salt

Tempeh—pressed soybean cake made from split soybeans, water, and special enzymes

Edamame—whole soybeans in a shell or without, kept frozen

Natto—whole cooked soybeans fermented with heart-healthy enzymes and whole grains

Sea Vegetables

Kombu

Arame

Hiziki

Nori—flakes or whole sheets

Dulse

Wakame

Leafy Green Vegetables

Asparagus

Beet greens

Broccoli

Cabbage

Carrot tops

Chinese cabbage

Collard greens

Curly dock

- Endive
- Escarole
- Kale
- Kohlrabi
- Lamb's quarters
- Leek greens
- Lettuce
- Mustard greens
- Parsley
- Plantain
- Scallions
- Shepherd's purse
- Sorrel
- Sprouts
- Swiss chard
- Turnip greens
- Watercress

Round Vegetables, Tubers, and Squash

- Artichokes
- Bamboo shoots
- Beets
- Brussels sprouts
- Green peas
- Mushrooms
  - Shiitake
  - Button
- Onions
- Squashes
  - Acorn squash
  - Butternut squash

- Hokkaido pumpkin
- Hubbard squash
- Pumpkin
- Yellow squash
- Zucchini

Snow peas
String beans
Sweet potatoes
Yams

Root Vegetables

Burdock
Carrots
Celery
Chicory root
Daikon radishes
Dandelion root
Icicle radishes
Jinenjo potatoes
Lotus root
Parsnip
Red radishes
Rutabaga
Salsify root
Turnips

Fruit

Apples
Apricots
Bananas

Blueberries

Blackberries

Boysenberries

Cherries

Cranberries

Cantaloupe

Grapes—red, green, and purple

Honeydew

Kiwi

Mango

Nectarines

Oranges

Papayas

Peaches

Pears

Plums

Prunes

Raisins

Strawberries

Tangerines

Raspberries

Watermelon

Fruit Spreads

Avoid spreads with added sugars.

Apple butter

Strawberry jam

Four fruits

Marmalade

Condiments

Tamari soy sauce

Sea salt

Umeboshi vinegar—salty sour pickled plum vinegar which originated in Japan, alkalizing, good salt substitute

Ume plum

Balsamic vinegar

Natural organic mustard

Organic sauerkraut

Fresh ginger

Fresh garlic

Miso (unpasteurized)

Brag's liquid aminos

Seeds and Nuts

Almond—tamari roasted almond

Dried chestnut

Sesame seeds

Sunflower seeds

Poppy seeds

Pumpkin seeds

Butters

Toasted sesame seed butter (Tohum's tahini is universally worshipped.)

Unsweetend apple butter

Sweeteners

Brown rice syrup

Dried unsulphured fruits such as raisins and apricots

Maple syrup

Raw honey

Beverages

Chamomile tea

Dandelion tea

Jasmine tea

Kukicha tea/green twig tea

Mu tea (very unique Chinese 16-herb mixture)

Mugicha tea/roasted barley tea

Nettle tea

Peppermint tea

Spring water

Snacks

Amasake—sweet, velvety smooth fermented rice beverage, comes in a variety of flavors

Baked brown rice crackers with sesame seeds, seaweed, and tamari soy sauce

Kamut cakes

Mochi—a traditional Japanese sweet brown rice cake

Toasted seeds

Rice syrup candies

appendix B
Where to Shop

Callie's Organics
1561 Post Road
Fairfield, CT 06430
203-255-1796
www.calliesfarm.com
Fresh produce online

Eden Foods
701 Tecumseh Road
Clinton, MI 49236
1-888-441-3336 or 1-888-424-3336
www.edenfoods.com
Full line of natural and organic foods

Diamond Organics
PO Box 2159
Freedom, CA 95019
1-888-674-2642
www.diamondorganics.com
Kitchen equipment and organic produce, grains, fruits, and other foods, shipped fresh overnight (guaranteed)

Join a CSA (Community Supported Agriculture) where you buy a share in an organic farm and have organic vegetables delivered to a location in your neighborhood every week of the growing season. Call 1-800-516-7797 or visit www.reeusda.gov/csa.htm for more information.

Jaffe Brothers
28560 Lilac Road
Valley Center, CA 92032
760-749-1133
Fax: 760-749-1282
www.organicfruitsandnuts.com
Organic fruits and nuts

Gold Mine Natural Food Co.
7805 Anjons Drive
San Diego, CA 92126
1-800-475-3663
www.goldminenaturalfood.com
Full line of organic and natural foods

Natural Lifestyle
16 Lookout Drive
Asheville, NC 28804
1-800-752-2772
www.natural-lifestyle.com
Full line of organic and natural foods

Miracle Exclusives
64 Seaview Boulevard
Port Washington, NY 11050
1-800-645-6360
www.miracleexclusives.com
Kitchen equipment

appendix C
A Guide to Eating Out

The following is a guide to making healthful food choices in restaurants. In fact, it's not that difficult, provided you follow some simple guidelines, which I describe later in this appendix. I have also given some suggestions for specific types of restaurants. First, here is some general advice.

No matter where you eat out, I suggest the following:

- Avoid red meat and instead choose fish, seafood, or skinless chicken.
- If you choose to eat red meat, choose the leanest cut and limit the serving size to 3½ ounces.
- Ask that your fish be broiled or steamed, rather than fried or cooked in butter or oil. If you choose chicken, ask that it be broiled without any oil or butter.
- Always order vegetables and ask that they be steamed in water or sautéed in olive oil. Avoid all other oils in restaurants. Vegetables are rich in nutrition, they'll fill you up, and they're low in calories.

- Order salad with at least two (preferably three) different vegetables. As a dressing, order olive oil, balsamic vinegar, and/or fresh lemon wedge on the side. Use the former two sparingly.
- No matter what you order, request that your food be cooked without butter. Avoid the Parmesan cheese in Italian restaurants. Both of these suggestions will cut down on the calories you consume, and help you dodge a bullet to your immune system.
- In Mexican restaurants, make sure your beans are not refried in lard. Most Mexican restaurants offer boiled beans that have no lard or oil added.
- Request low-sodium soy sauce or shoyu in Chinese and Japanese restaurants.
- Avoid the bread in restaurants. It will add lots of additional calories and fill you up before the food that you really want arrives.
- Order vegetable soup, such as minestrone, whenever it's available. It's low in calories, nutritious, delicious, and filling.
- As much as possible, choose your restaurants for quality, not price. All too often, you walk out of a greasy spoon feeling that you damaged your health and wasted your money. It's not worth it. Spend a few dollars more and eat something that will make you feel good—not just while you're in the restaurant, but after you leave, too.
- Even when you choose a fast-food restaurant, order fresh vegetables and include salad. You'll get more nutrition, and the fiber will help you digest and eliminate any fat you may have consumed.

Whether you are traveling to an unfamiliar city or just going out to your local restaurant, you can always get a healthful meal that is rich in nutrition, relatively low in calories, and completely satisfying. Most American restaurants provide delicious fish dishes, steamed or boiled vegetables, and salads. Many American restaurants also provide bean dishes and vegetable soups. Japanese, Chinese, and other Asian restaurants are widely available, and all of them provide extremely healthful and low-calorie choices. Italian restaurants provide delicious pasta, vegetables, fish, seafood, and soups. Mexican restaurants offer rice, beans, vegetables, and steamed tortillas. Even many French restaurants provide nouvelle cuisine, which is composed primarily of healthful vegetables and low-fat animal foods, prepared to perfection.

There's no reason why you cannot go out and enjoy yourself, even as you maintain a healthful and weight-reducing diet. Here are some suggestions for restaurants and for healthful food choices.

American Restaurants

Any combination of the following are healthful choices:

- Vegetable soup (Soups are often made with chicken broth. If you don't mind the chicken, fine, but if you do, ask your waitperson so you can avoid the chicken—and the fat.)
- Salad, with olive oil, balsamic vinegar, or lemon-wedge dressing
- Steamed vegetables of the day
- Fish, broiled or steamed
- Baked potato, with chives and other vegetables, but without sour cream or butter
- Tea, juice, beer, or beverage of choice
- Fruit dessert

Japanese Restaurants

Some healthful choices:

- Miso soup
- Edamame (steamed soybeans, lightly salted) as an appetizer
- Seaweed salad
- Green salad
- Vegetables (Japanese restaurants offer a wide variety of vegetable dishes, made without oil or butter. Ask your waitperson.)
- Rice (Many Japanese and Chinese restaurants now offer brown rice. Ask if it's available.)
- Fish or vegetable sushi
- Steamed fish dishes
- Sake, tea, or beverage of choice
- Fruit dessert

Chinese Restaurants

Some healthful choices include the following:

- Steamed vegetable appetizer
- Steamed rice (Request brown rice, if available.)
- A wide variety of fish and vegetable dishes are available, including:
  - Shrimp and vegetables
  - Steamed whole fish and vegetables
  - Scallops and vegetables
  - Noodle dishes and vegetables
  - Rice noodle dishes, for people avoiding wheat
  - Stir-fried chicken and vegetables
- Tea or beverage of choice
- Fortune cookie

Italian Restaurants

Any of the following are healthful choices. If anything is prepared in olive oil, ask the cook to go light on the oil.

- Salad, with olive oil, balsamic vinegar, and/or lemon wedge
- Broccoli rabé, steamed or lightly sautéed in olive oil
- Asparagus or other vegetables of the day, steamed or lightly sautéed in olive oil
- Linguine or spaghetti primavera (pasta cooked with vegetables) in marinara sauce
- Linguine with red clam sauce (cooked without butter)
- Fruit d' mare with linguine or spaghetti or some other pasta (seafood pasta)
- Fish of the day, steamed or broiled
- Glass of wine or Italian beer
- Tea
- Sorbet or fruit dessert

Mexican Restaurants

Some healthful choices include the following:

- Salad of green vegetables and tomatoes
- Black bean burrito (cooked without oil or lard)
- Steamed corn tortilla, beans, and vegetables
- Tacos, beans, and vegetables
- Salsa
- Guacamole (avocado and tomatoes)
- Black beans and rice
- Spring water or beverage of choice

appendix D

Comprehensive List of Contracting and Expanding Foods

The following is a guide to the expanding and contracting effects of foods. The first two lists are for extreme foods only; the first for extreme contracting foods and the second for extreme expanding ones. Within each list, I have arranged the foods to indicate their relative degree of either contraction or expansion. For example, in the extreme contracting list, the first food is salt, the most contracting substance in the food supply, while the last food is the fish and shellfish, the least contracting, or most balanced. The second list is for extreme expanding foods. Within that list, the foods are arranged from least expanding, beginning with white bread, to most expanding, namely hard liquor.

As you will see, the foods that make up most diets of people today are those in the extreme expanding and extreme contracting categories. There are a limited number of those foods.

Those lists are followed by the mildly contracting and mildly expanding foods. I have tried to rate the foods within these lists according to their relative effects on the body, as well. Also, the lists themselves range from most contracting (the grain list) to least contracting (the fermented foods and condiments list), and from least expanding (the root vegetables list) to most expanding (the fruit and fruit juices list).

There are three lists within the mildly contracting category—namely, the whole grains, the sea vegetables, and the fermented foods and condiments. The grains are more contracting than the sea vegetables and the sea vegetables more contracting than the fermented foods and condiments. The reason sea vegetables and fermented foods have a contracting effect on the body is the presence of salt.

As I explained in Chapter 1, foods that have an extreme contracting effect on the body create physical tension, digestive discomfort, bloat, constipation, fatigue, sleepiness, limitations in range of motion, and cravings for extreme expanding foods, such as sugar, pastries, and alcohol. When we eat extreme contracting foods, we often feel trapped in the body, blocked, and stagnant.

These feelings are often relieved by extreme expanding foods, which provide an almost immediate rush of energy, relief in digestion, increased circulation, and a lighter feeling in the body. When eaten consistently, however, these foods create low blood sugar, fatigue, weakness, nervous tension, anxiety, and cravings for nutrition and extreme contracting foods.

When the diet is dominated by extreme contracting and extreme expanding foods, we are driven from one set of cravings to the next. We have great difficulty experiencing any degree of balance, self-control, or freedom, especially from food.

Mildly contracting foods provide feelings of strength, power, centeredness, stability, and vitality, but do not cause the severe cravings for extreme expanding foods. Mildly expansive foods create feelings of energy, enhanced digestion, and lightness of being, without creating severe cravings for extreme contracting foods. When the diet is dominated by mildly contracting and mildly expanding foods, we experience balance, clarity of thought, physical ease, reduction of symptoms, deeper, more restful sleep, freedom from severe cravings, dramatically improved health, and increased energy levels.

Here are the representative lists of extreme contracting and extreme expanding foods, followed by mildly contracting and mildly expanding foods.

Extreme Contracting Foods

(Most contracting to least contracting.)

Salt
Beef
Pork, ham
Venison
Lamb
Chicken
Turkey
Eggs
Hard cheeses
Fish
Shellfish

Extreme Expanding Foods

(Least expansive to most expansive.)

Bagels
White bread
White rolls
Pastries
Doughnuts
Sugar
Foods that contain artificial ingredients
Cow's milk
Coffee and other caffeinated beverages
Chocolate
Candy
Jell-O and puddings
Soft drinks
Beer

Wine
Hard liquors

Mildly Contracting Foods

(Most contracting to least contracting.)

Whole Grains

Brown rice
Sweet rice
Mochi
Millet
Barley
Buckwheat
Whole wheat
Teff
Amaranth
Quinoa
Oats
Spelt
Buckwheat pastas, such as Japanese soba
Whole-wheat pastas
Sifted wheat pastas, such as Italian pastas and Japanese don
Rice noodles

Sea Vegetables

Alaria
Arame
Dulse
Hijiki
Kombu
Nori
Wakame

Fermented Foods and Condiments

Miso
Tamari
Shoyu
Daikon pickle
Umeboshi plum
Sauerkraut
Dill pickles

Beans

Aduki beans
Black beans
Black-eyed peas
Chickpeas
Kidney beans
Lima beans
Lentils
Navy beans
Pinto beans
Soybeans
Tempeh
Natto
Edamame
Tofu
Split peas

Mildly Expanding Foods

(Least expanding to most expanding.)

Vegetables

Roots

- Burdock
- Carrots
- Chicory root
- Daikon radishes
- Dandelion root
- Ginger root
- Parsnips
- Red radishes
- Rutabagas
- Turnips
- Icicle radishes
- Jinenjo potatoes
- Lotus root

Round and Close-to-the-Ground Vegetables

- Artichokes
- Bamboo shoots
- Beets
- Cucumbers
- Green peas
- Leeks
- Shiitake mushrooms
- Okra
- Squashes:
  - Acorn squash

- Butternut squash
- Hokkaido pumpkin
- Hubbard squash
- Pumpkin
- Yellow squash
- Zucchini squash

Snow peas
String beans
Sweet potatoes
Yams
Onions
Green and leafy vegetables
Asparagus
Beet greens
Carrot tops
Chinese cabbage
Broccoli
Brussels sprouts
Cabbage
Collard greens
Curly dock
Daikon radishes
Dandelion greens
Endive
Escarole
Kale
Kohlrabi
Lamb's quarters
Leek greens
Lettuce
Mustard greens

Parsley
Plantain
Scallions
Shepherd's purse
Sorrel
Sprouts
Swiss chard
Turnip greens
Watercress

Fruit

Apples
Apricots
Bananas
Blueberries
Blackberries
Boysenberries
Cherries
Cranberries
Cantaloupe
Grapes: red, green, and purple
Honeydew
Kiwi
Mango
Nectarines
Oranges
Papayas
Peaches
Pears
Plums

Prunes

Raisins

Strawberries

Tangerines

Raspberries

Watermelons

Dried Fruit

(Dried fruit is rich in sugar and calories, which makes them more expanding than fresh or raw fruit.)

Raisins

Dried apples

Dried apricots

Prunes

Dates

Figs

Fruit Spreads (No Sugar Added)

(Fruit spreads are more expanding than dried fruit because more sugars have been released and the fiber removed.)

Fruit preserves of all types

Apple butter

Strawberry jam

Four fruits

Marmalade

appendix E
Glossary of Health-Promoting Foods

Whole Grains and Grain Products

amaranth The grain of the Aztecs, amaranth is sometimes referred to as the miracle grain because of its high protein and nutrient content. In fact, its high quantities of calcium, folic acid, iron, magnesium, phosphorus, potassium, iron, and other nutrients have caused people to label amaranth as a super grain. In traditional Chinese medicine, amaranth is used to strengthen the heart and small intestine. It can be boiled with or without vegetables and ready in about 30 minutes.

barley Barley is one of the first grains cultivated by humans. From ancient times, people have used barley as a form of medicine and even as a form of currency. In traditional Chinese medicine, barley is used to strengthen the kidneys, bladder, and liver. Barley comes in many forms, from whole barley to "pearled" to flakes. Whole barley has had only the inedible outer hull removed and therefore is highly nutritious. It is a good source of B vitamins, iron, magnesium, manganese, niacin, phosphorous, potassium, riboflavin, thiamine, and zinc. Pearled barley has had both the outer and inner hulls removed. Though it has been reduced in nutritional value, it requires less time to cook. Whole barley takes about an hour and a half to boil and fully cook, while pearled barley requires about 55 minutes. Barley is worth the extended cooking time, however. Boil it with lots of mushrooms, carrots, and leeks, and then add miso or tamari after it's been fully cooked. Delicious and very good for your health.

brown rice Rice is the staple of food of China, Japan, and other Asian countries; it's also popular in Italy and North America. Brown rice is a good source of B vitamins, such as thiamine and niacin. It also provides iron, magnesium, and phosphorus. The soluble fiber in rice has been shown to lower cholesterol. Brown rice grows in three different sizes: short, medium, and long grain. Rice is used in traditional Chinese medicine to strengthen the lungs and large intestine. In addition to the three sizes of grains, brown rice also comes as sweet rice, which is much more glutinous. Sweet rice is pounded into patties to make mochi, a hearty and strengthening food that the Japanese traditionally eat at New Year's to promote greater strength for the year ahead. Mochi can be baked in an oven to make a chewy, puffed rice patty, or chopped up into squares and added to soup to make soft dumplings.

buckwheat Not a true grain, buckwheat is actually the fruit of a leafy plant that is related to rhubarb and sorrel. Buckwheat comes as whole groats (after the inedible hull has been removed); the groats are white when they haven't been roasted and brown when they have. When the whole groats are cracked and roasted, they become kasha, a traditional dish of Eastern Europeans and the peoples of northern Asia. Buckwheat has been one of the most popular grains in cold climates because of its strong capacity to warm the body. It has long been used traditionally to strengthen the kidneys and bladder. Buckwheat has a strong nutty flavor that is delicious when combined with vegetables (squash, roots, or greens) or cooked separately and then combined with bow-tie noodles. (Boil noodles, drain, add to a frying pan coated with olive or sesame oil, and then ladle in the buckwheat. Heat, stir, and enjoy.)

kamut Another ancient relative of wheat, but is far easier to digest, has less allergic potential, and provides more protein, copper, iron, potassium, and zinc than wheat.

millet Millet is a highly nutritious grain and a rich source of B vitamins, copper, and iron. Long a staple grain in North Africa, China, and India, millet has no gluten. Millet is prepared in under a half-hour. Boil it with cauliflower and other vegetables or make it as part of a soup. Delicious and highly strengthening.

oats The water-soluble fiber in oats lowers cholesterol, as many of the grains do. Oats have about 50 percent more protein than wheat, and twice as much as brown rice. Oats also provide copper, folic acid, vitamin E, and zinc. Oats come whole, steel cut, and rolled. Rolled oats are the most highly processed and the quickest cooking. Simply bring them to a boil and simmer for five to eight minutes and they are ready.

quinoa Not a true grain, but like amaranth and buckwheat it is a grass. Quinoa contains high levels of potassium, riboflavin, magnesium, manganese, zinc, copper, and folic acid. It is also an excellent source of protein. Quinoa can be boiled and served in 15 minutes. Cook with vegetables, herbs, and spices for a delicious and easy-to-prepare grain.

spelt An ancient relative of wheat, spelt is easier to digest than wheat, has low gluten, and is a good source of protein, copper, zinc, and B vitamins.

teff A small, delicious grain—it has a molasseslike flavor—that's rich in calcium and low in gluten.

triticale A hybrid grain cultivated in the nineteenth century by a Scottish botanist who crossed wheat and rye. It is glutinous and a good source of B vitamins, folic acid, magnesium, phosphorus, and potassium.

whole wheat Wheat is one of the first grains cultivated by humans and has been eaten throughout Asia, Europe, North Africa, and America. Wheat is a good source of protein, B vitamins, iron, magnesium, manganese, many other minerals and vitamins. There are many forms of wheat, including Hard Red Winter, Hard Red Spring, Soft Red Winter, Hard White, Soft White, and Durum wheat. Most people get their wheat from bread, pasta, and other flour products, but wheat is best eaten as a whole grain, namely as wheat berries, and as bulgar, which is the cracked grain. Whole wheat berries are delicious, but require much more time to prepare—about an hour and 10 minutes. Bulgar wheat can be boiled and served in 15 minutes.

Among the many products made from wheat are fu, which is baked, puffed wheat gluten; and seitan, also known as "wheat meat." Seitan is meaty, hearty, and glutinous. It is made from the protein in the wheat and often used in soups, stews, and vegetable medleys.

Beans and Bean Products

aduki Highly regarded in Japan for their strengthening and medicinal qualities—especially for promoting kidney strength—aduki beans are small, red beans that contain the least amount of fat of all the beans. They can be cooked alone or combined with a grain, such as brown rice or sweet rice, to make a rich, hearty dish.

black beans A staple of Latin America, black beans (also known as black turtle beans) are used in traditional Chinese and Japanese medicine as a health-promoting food for the kidneys, bladder, and reproductive organs.

black-eyed peas Long a traditional delicacy of the Southern United States and commonly used in salads.

cannelloni Commonly used in Italian cooking, cannelloni beans are large and white and usually used in minestrone soup.

chickpeas A staple of the Middle East and the Mediterranean, chickpeas are also called garbanzos. More and more supermarkets are carrying organic chickpeas that have been precooked and jarred. Just take them home and reheat and you've got a delicious bean. Use them in salads or as part of vegetable medleys.

fava beans Another bean favored by Mediterranean peoples, especially Italians. Cook with mild spices and a dash of pepper.

kidney beans A common bean in soups and, along with pinto beans, the most commonly used beans in Mexican cooking.

lentils The quickest cooking beans—red lentils take about 20 minutes to prepare, green about 35—and very easy to digest.

lima beans Rich and luscious, lima beans are wonderful when combined with squash, corn, and other vegetables to produce succotash and other bean-and-vegetable medleys.

mung beans A small green or black bean with a sweet, hearty taste.

navy beans The smaller cousin of the Great Northern bean, navys are great in stews and soups. Very hearty and warming.

pinto beans Would Texas be Texas without pinto beans and chili? Not hardly, pilgrim. Wonderful when combined with carrots and onions.

soybeans By themselves, soybeans are tough, hard to digest, and take a long time to prepare. This alone might make people avoid them, but the

Japanese and Chinese saw how nutritionally rich this food really was and so came up with a whole array of foods that stemmed from this otherwise stubborn bean. Among the things they did was to create the following foods, all of which are bursting with health-promoting phytochemicals, including those all-important plant-based estrogens.

- **Edamame** The whole soybean, precooked and available in the freezer sections of most supermarkets. All you have to do is steam and pop the beans out of the shells. Sprinkle some salt lightly on the beans. A wonderful snack.
- **Miso** A thick, rich paste made from aged and fermented soybeans that is used as a base for soups, stews, and sauces. Miso, a traditional food of Japan, is abundant in health-promoting digestive enzymes and bacteria. Therefore, it should not be boiled—that would only kill the bacteria in the food; instead, add miso to the dish after it has been fully cooked and the flame has been turned off. Miso is also a rich source of the phytochemical, genistein, which blocks blood flow to tumors and thus deprives them of oxygen and nutrients. There are many misos. The differences among them are the length of time they were aged and the presence of other beans and grains used in the fermentation process. Among the most popular forms of miso are those made with rice, barley, chickpeas, and millet.
- **Natto** Whole soybeans that have been cooked and fermented to make a condiment for grains and noodles.
- **Shoyu** Aged and fermented soy sauce that contains no artificial colors, flavors, or chemicals designed to speed-up the fermentation process. Shoyu provides digestive enzymes and health-promoting bacteria. Shoyu contains sodium and some shoyus can contain as much as 18 percent sodium. Low-sodium shoyus are widely available and are just as delicious as those that are richer in sodium.
- **Soymilk and various soy cheeses** Interestingly enough, soymilk is a rich supplier of isoflavones; it also contains about 150 mg of calcium. But it's highly processed and should be used in moderation.
- **Tamari** The liquid-run-off produced during the miso-making process is called tamari. It is jarred and used as a base for soups, stews, and sauces. Tamari is used in cooking just like shoyu, or soy sauce, and is extremely flavorful and health-enhancing. Tamari supplies friendly flora and digestive enzymes.

- **Tempeh** Cooked and fermented soybeans that have been pressed into a patty, tempeh offers a rich supply of health-promoting bacteria, enzymes, protein, calcium, and phytoestrogens. Place tempeh in soups and stews and boil, or fry it in sesame oil until brown. Fried tempeh is delicious and can be added to soups and noodle broths.
- **Tofu** Like all soybean products, tofu provides a rich supply of the phytoestrogen genistein, which has been shown to help cut off blood supply to tumors, thus starving them of blood and oxygen. Tofu is also a great source of calcium (4 ounces provide 150 mg). Tofu is wonderful when added to soups, stews, and noodle broths. It can be baked and eaten raw (with a little shoyu and ginger added), as well. It is not fermented.

split peas A wonderful stock in soups and stews. Like lentils, they are quick cooking—they need only 30 to 35 minutes to prepare—and are delicious and nutritious.

Sea Vegetables or Seaweeds

agar agar A thickening agent that usually comes in translucent flakes or bars. It is flavorless and odorless. When added to warmed fruit juice or sauces, agar agar creates a gelatinlike consistency.

arame A delicate, grasslike sea vegetable that is delicious when cooked with other vegetables, such as carrots and onions, in lemon juice. Arame is abundant in vitamins A and B, carbohydrate, calcium, many other minerals, and trace elements.

dulse Usually added to soups and stews, or roasted and then broken into tiny pieces and added to grain as a condiment. It's a good source of protein, vitamins A, C, E, and B vitamins, iodine, minerals, and trace elements.

hijiki or **hiziki** A stringlike seaweed, with a strong, salty flavor, hijiki is one of the most nutritious foods on the planet. Three-and-a-half ounces provide 1,400 mg of calcium. But hijiki is also a rich source of protein, vitamins A, and B complex family, phosphorous, iron, and many trace elements. Boil with carrots, onions, and daikon radish for 1 hour to 1½ hours.

kombu One of those seaweeds that can be added to any soup, bean, grain, or noodle broth to boost the nutritional value without influencing the taste. Kombu comes in stalks. Add a stalk to just about anything you are cooking and you will boost its immune-strengthening and cancer-fighting abilities dramatically.

nori Nori comes in pressed sheets. Regular nori has to be roasted—just hold it over an open flame for a few seconds and it crinkles up and is ready to be eaten. Sushi nori has already been roasted and can be eaten directly from the package. Nori is the seaweed that is wrapped around sushi and sushi rolls. There are other forms of nori that are flavored and spicy. These are delicious; children enjoy them as a snack. Nori is another nutrient-rich sea vegetable, rich in vitamins A, the B complex family, C, and D, calcium, phosphorous, iron, and trace elements.

wakame Used most often in soups (especially miso soup), stews, and noodle broths, wakame is highly nutritious, super rich in calcium (1,300 mg per 3½-ounce serving), vitamin A, B vitamins, and potassium. Cooks in 20 minutes.

Thickening Agents

agar agar A clear, crystaline seaweed that is used in cooking as a thickening agent to create gelatins and aspics.

kuzu or **kudzu** A white, powdery starch that usually appears as rocks, kuzu comes from the root of the wild kudzu plant. It is used to thicken liquids, including desserts, soups, and sauces. It is highly alkalizing and used traditionally to strengthen and heal the intestinal tract. It can be very healing for all digestive problems.

Index

A

B

C

E

F

H

I

S

U–V

W–X–Y–Z

About the Authors

Joshua Rosenthal is the founder and director of the Institute for Integrative Nutrition located in New York City. He is the former owner and operator of 12 Physicians Weight Loss clinics and 4 natural foods restaurants. He was licensed by the Canadian Government's Ministry of Health as a drugless practitioner. He is also the founder of *Common Ground*. Joshua has a Master's of Science degree in education and counseling from Duquesne University and maintains a private counseling practice for health and natural healing in New York City. He has lectured in the United States, Canada, England, Europe, and Australia.

Tom Monte is a best-selling health and science writer. He has written 26 books on health and healing, including *The Complete Guide to Natural Healing; World Medicine: The East West Guide to Healing Your Body; The Ten Best Tools to Boost Your Immune System* with immunologist Elinor Levy, Ph.D.; *Staying Young; The Way of Hope;* and *The McDougall Program for Healthy Women* with John A. McDougall, M.D. Articles by Tom Monte have appeared in *Life, The Saturday Evening Post, Cosmopolitan, Runner's World, The Chicago Tribune, East West Journal, New Age Journal, Natural Health,* and *Ladies' Home Journal.*

The Institute for Integrative Nutrition

The Institute for Integrative Nutrition is among the most exciting teaching centers in the world for the study of nutrition and the use of diet as a tool for personal and social transformation. Our students are trained to become professional counselors and teachers in the areas of nutrition, diet, and health.

Among the most important themes of our school is to integrate the many diverse theories and approaches to diet and healing. Our students are taught from both the modern and traditional perspectives. We teach nutrition science and the theories that underlie the most popular dietary practices today—everything from raw foods diets to high-carbohydrate, low-fat programs to the high-protein, low-carbohydrate regimens. We thoroughly explore the dietary and healing practices of ancient traditions, such as Chinese medicine and Ayurveda.

Our students emerge with a truly holistic understanding of diet and healing. They know the science and the complementary systems people are so attracted to today. Even more exciting, our graduates are able to integrate these diverse systems to create successful healing programs that meet the unique needs of their individual clients.

It is our belief that there is no one diet for all humans. Life is always changing. One of the ways we adapt to change is by modifying our dietary choices to meet our ever-changing nutritional and psychological needs. We teach students how they can do that and, at the same time, promote their own good health, clarity of mind, and sound judgment.

In addition to diet and nutrition, our faculty presents information on a wide array of other transformational healing modalities, such as various forms of bodywork, physical therapies, and mind-body techniques for healing.

Our students emerge from our school with a fundamental and practical knowledge of East and West, of the traditional and modern approaches to healing, and, most important, a profound knowledge of themselves. We believe that with the knowledge our students receive at the Institute for Integrative Nutrition, they are the best-prepared counselors for helping people achieve a lifetime of good health and powerful lives.

The Institute for Integrative Nutrition is located at 120 West 41st Street, 2nd Floor, New York, NY 10036 (phone: 212-730-5433). Check our website at www.integrativeNutrition.com or e-mail us at info@integrativeNutrition.com.

You can learn more about the Energy Balance Diet at www.energybalancediet.com.